COMMUNITY ECOLOGY OF NEOTROPICAL KINGFISHERS

Community Ecology of Neotropical Kingfishers

J. V. Remsen, Jr.

UNIVERSITY OF CALIFORNIA PRESS
Berkeley • Los Angeles • Oxford

UNIVERSITY OF CALIFORNIA PUBLICATIONS IN ZOOLOGY

Editorial Board: Peter B. Moyle, James L. Patton,
Donald C. Potts, David S. Woodruff

Volume 124
Issue Date: December 1990

UNIVERSITY OF CALIFORNIA PRESS
BERKELEY AND LOS ANGELES, CALIFORNIA

UNIVERSITY OF CALIFORNIA PRESS, LTD.
OXFORD, ENGLAND

© 1991 BY THE REGENTS OF THE UNIVERSITY OF CALIFORNIA
PRINTED IN THE UNITED STATES OF AMERICA

Library of Congress Cataloging-in-Publication Data

Remsen, James Vanderbeek
    Community ecology of neotropical kingfishers: by J.V. Remsen, Jr.
       p.   cm. — (University of California publications in zoology;
v. 124)
    Includes bibliographical references.
    ISBN 0-520-09673-8
    1. Kingfishers—Latin America—Ecology. 2. Kingfishers—South
America—Ecology. 3. Kingfishers—Latin America—Feeding and feeds.
4. Kingfishers—South America—Feeding and feeds. 5. Kingfishers—
North America—Ecology. 6. Kingfishers—North America—Feeding and
feeds. 7. Bird populations—Latin America. I. Title. II. Series.
QL696.C72R45   1991
598.8'92—dc20   90-46525
                                                               CIP

Contents

List of Figures, vii
List of Tables, viii
Acknowledgments, ix
Abstract, xi

INTRODUCTION   1
    New World Kingfisher Distribution, 4

STUDY SITES   10

FISH DATA   13
    Sampling Methods, 13
    Fish Densities, 14
    Size-class Proportions, 18
    Taxonomic Composition, 20

KINGFISHER DENSITIES   24
    Territorial Behavior, 24
    Density Sampling Methods, 26
    Seasonal Density Differences, 28
    Densities in Different Habitats, 32
        Lakes, 32
        Streams, 34
        Rivers, 38
        Differences Between Rivers and Streams, 38
    Interaction of Perch Availability and Fish Availability, 40

KINGFISHER FEEDING BEHAVIOR 44
    Study Methods, 44
    Qualitative Description of Feeding Behavior, 45
    Habitat Preferences, 46
    Perch Type, 48
    Perch Height, 50
    Horizontal Location of Perches, 53
    Dive Entry Point, 58
    Prey Size, 60
    Prey Type, 64

OTHER PISCIVOROUS BIRDS 68
    Osprey *(Pandion haliaetus)*, 68
    Large-billed Tern *(Phaetusa simplex)*, 69
    Yellow-billed Tern *(Sterna superciliaris)*, 70
    Black-collared Hawk *(Busarellus nigricollis)*, 73
    Lesser Kiskadee *(Pitangus lictor)*, 74
    Green-backed Heron *(Butorides striatus)*, 75
    Great Egret *(Casmerodius albus)*, 76
    Summary, 76

SPECIES-PACKING MECHANISMS 78
    Niche-Breadth Mechanism, 78
    Niche-Overlap Mechanism, 81
    Resource-Base Mechanism, 89

DISCUSSION 94
    Assumptions and Conclusions, 94
    Hypotheses Concerning Tropical Species Diversity, 96

Literature Cited, 101

List of Figures

1. Kingfisher species density in North America, 5
2. Kingfisher species density in South America, 6
3. Water-level fluctuations of the Amazon River at Isla de Santa Sofía, Amazonas, Colombia, 11
4. Patterns of seasonal change in fish density in different size classes in three habitats, 17
5. Proportion of *Cichlasoma festivum* in surface fish censuses in various size classes for three habitats, 22
6. Comparison of degree of seasonal change in kingfisher densities with degree of seasonal change in prey availability for each kingfisher, 30
7. Relationship between surface fish density and bird density for each study site at low-water season, 37
8. Relationship between surface fish density and bird density at high-water season, 41
9. Relationship between availability of completely shaded perches and density of *Chloroceryle inda* and *C. aenea,* 42
10. Perch heights in five-species kingfisher community, 51
11. Perch heights in the three-species kingfisher community, 52
12. Relationship between horizontal location of perch in terms of distance from effective shore and height of perch for *Chloroceryle torquata* in the five-species community, 55
13. Relationship between horizontal location of perch in terms of distance from effective shore and height of perch for *Chloroceryle amazona* in the five-species community, 56
14. Relationship between horizontal location of perch in terms of distance from effective shore and height of perch for *Chloroceryle americana* in the five-species community, 57
15. Lengths of kingfisher prey in the five- and three-species communities, 62
16. Tern hunting heights, 71
17. Hunting distances from shore for terns, 72
18. Schematic representation of the niche-breadth mechanism, 79
19. Schematic representation of the niche-overlap mechanism, 82
20. Schematic representation of the resource-base mechanism, 90
21. Kingfisher diet vs. surface fish densities for various fish size-classes, 91

List of Tables

1. Bill lengths of adult kingfishers, 7
2. Surface fish densities at the intensive study sites, 15
3. Proportions of surface fish population in size-classes, 19
4. Proportions of Characidae in surface fish populations, 21
5. Number of aggressive interactions observed between kingfisher species, 25
6. Availability of kingfisher hunting perches, 27
7. Ratios of low-water to high-water season densities of kingfishers, 28
8. Densities of piscivorous birds at lake sites, 33
9. Densities of piscivorous birds at stream sites, 35
10. Densities of piscivorous birds on rivers, 39
11. Kingfisher habitat preferences, 47
12. Use of perch types by kingfishers, 49
13. Distance from shoreline of observed kingfisher hunting perches, 54
14. Distances from shore for kingfisher dive entry points, 59
15. Prey types taken by kingfishers, 65
16. Kingfisher niche breadths, 80
17. Overall niche-overlap values using multiplicative index, 85
18. Overall niche-overlap values using additive index, 86
19. Bill lengths of piscivorous kingfishers of the Amazon and Congo basins, 97

Acknowledgments

This study was conducted under the guidance of Dr. Frank A. Pitelka. Field research was supported by a Doctoral Dissertation Grant from the National Science Foundation and by the Frank M. Chapman Memorial Fund of the American Museum of Natural History. Computer funds were provided by the Department of Zoology, University of California, Berkeley, through the assistance of Dr. Pitelka, and by the Computer Center, University of California, Berkeley. I am grateful to Ronn Storro-Patterson and Fernando Ortiz-Crespo for arranging for my first visit to the Amazon. Dennis E. Breedlove first pointed out to me the potential of the kingfisher problem.

Dr. Pitelka's comments on an early draft of the manuscript were extremely useful, as were those of Robert K. Colwell, Russell Greenberg, Judy Gradwohl, and Robert Ornduff. Later drafts benefited greatly from the careful reviews of David E. Willard, George L. Hunt, James L. Patton, and an anonymous reviewer.

I thank the following individuals for valuable discussions and comments concerning various aspects of kingfisher biology: Stephen F. Bailey, Alan Bond, Michael Christie, R. Glenn Ford, C. H. Fry, Ned K. Johnson, Walter D. Koenig, Paul Loiselle, James Lovisek, J. P. Myers, T. A. Parker, H-U. Reyer, K. V. Rosenberg, and Stephen D. West. I am grateful to the following persons for providing information on kingfisher distribution and behavior: John W. Fitzpatrick, Thomas R. Howell, James R. Karr, John P. O'Neill, T. A. Parker, III, Alexander F. Skutch, Paul Slud, Samuel Sweet, David E. Willard, and Edwin O. Willis.

Field research on Isla de Santa Sofía II in Colombia was made possible through the generosity and interest of Mike Tsalickis of Leticia, Amazonas, Colombia. Edelson Brito and Robert C. Bailey provided much needed information and aid in Colombia. Permission to do research and collect specimens in Colombia was granted by Jorge Hernandez and Victor Hugh Vasquez V.

Field research at Tumi Chucua, Bolivia, was made possible through the generosity of the Summer Institute of Linguistics, particularly Bob Wilkinson and Ron Olson. I am extremely grateful for the overwhelming hospitality and help at Tumi Chucua from Bob and Lois Wilkinson, Guy and Jean East, Ron Williams, and many others too numerous to mention. Tom Hutson enabled me to find a suitable research site in the Bolivian pampas. Jaime Cuellar and Ester de Cuellar generously extended to me the facilities of Estancia Inglaterra on the Río Yata. Encouragement and permission to work in Bolivia were granted by Prof. Gastón Bejarano, and initial help was given by David L. Pearson.

I thank the following persons and institutions for permission to examine specimens in their care: Laurence C. Binford (California Academy of Sciences); John W. Fitzpatrick and D. E. Willard (Field Museum of Natural History); Frank B. Gill and Mark B. Robbins (Academy of Natural Science, Philadelphia); Thomas R. Howell (University of California, Los Angeles); Joseph R. Jehl, Jr. (San Diego Natural History Museum); Ned K. Johnson (Museum of Vertebrate Zoology); Lloyd Kiff (Western Foundation of Vertebrate Zoology); Wesley E. Lanyon and John Farrand (American Museum of Natural History); Antonio Olivares and R. Romero Z. (Universidad Nacional, Bogotá); John P. O'Neill (Museum of Zoology, Louisiana State University); Ralph W. Schreiber (Los Angeles County Museum); and Betsy Webb (Denver Museum of Natural History).

Robert E. Jones provided much needed assistance in the preparation of field equipment and supplies. Dustin D. Chivers, Al Smalley, and Warren C. Freihofer provided preliminary identifications of invertebrates and fishes. Karen L. Bailey, Alice J. Fogg, Susan Jankus, Gail Kinney, Amy Parrish, and Musette M. Richard expertly typed various drafts or portions of drafts of the manuscripts, and Jana Kloss and Rose Anne White helped with preparation of the camera-ready copy.

Abstract

The purpose of this study was to investigate the factors responsible for the latitudinal gradient in species diversity in Neotropical kingfishers. Feeding ecology, bird density, and prey availability were quantified at three study sites in South America. The first site, which included a series of lakes, an island in a major river, and a forested stream, was about 50 km northwest of Leticia, Amazonas, Colombia, on the Amazon River; this site is near the geographic center of the region of maximum species density (5 species: *Ceryle torquata*, *Chloroceryle amazona*, *C. inda*, *C. americana*, and *C. aenea*). The second site, a large oxbow lake near Riberalta, Dpto. Beni, Bolivia, is near the southern limit of the region with five sympatric species. Most parameters of kingfisher feeding ecology showed no significant differences between these two sites. The third site, a small stream flowing into the Río Yata in the savannas of Bolivia, about 250 km south of Riberalta, is near the northern extreme of the region with three sympatric species. Conclusions concerning the latitudinal gradient in diversity were based primarily on comparisons among the five- and three-species sites.

Surface fish density at the site with three kingfisher species was 1.3-18.3% of that at the sites with five species. Kingfisher density at nine subsites within the major sites was positively correlated with surface fish density, and this correlation was further improved by correction for differences in availability of kingfisher hunting perches. Degree of seasonal change in density for each kingfisher species was positively correlated with degree of seasonal change in fish density in the size-classes preyed upon by the kingfisher in its primary habitat.

Three kingfishers (*torquata*, *amazona*, and *americana*) frequented primarily open habitat along the edges of lakes and streams, whereas two species (*inda* and *aenea*) foraged in shaded habitat, such as flooded forest and small forest streams, to a much greater extent. Differences among the kingfishers in selection of perch types reflected differences in perch heights or habitats used rather than any ecological separation based on perch type per se. The larger the kingfisher, the higher above water was its mean perch height. Kingfishers at the three-species site hunted from significantly lower perch heights than at the five-species sites. All species at all sites concentrated dive entries in the first 2 m from shoreline; surface fish density was much higher within that zone than farther from shore.

Only aquatic prey were taken, either fishes or, rarely, invertebrates. Kingfishers did not seem to select actively any taxonomic category of prey. One species of fish, *Cichlasoma festivum*, which formed a substantial portion of the total population of surface fishes, was rarely if ever captured by kingfishers. Differences in selection of prey types among kingfishers were interpreted as a result of differences in prey-size selection and unequal distribution of prey types within the spectrum of prey sizes. Differences in prey-size selection were the primary means of resource partitioning. The bill length of each kingfisher species was strongly correlated with mean prey size for both the five- and three-species communities. However, two kingfisher species in the three-species community took significantly smaller fishes at the three-species site than at the five-species sites. This downward shift in prey-size selection was almost certainly the result of the greatly reduced fish density, most pronounced among larger fish size-classes. This change in surface fish populations in open habitats, combined with extremely low fish density in shaded habitats — evidently so low that foraging there was completely unprofitable — was proposed to be the proximate cause for the change from five to three kingfisher species. Ecological separation of kingfishers from other piscivorous birds was also discussed.

The kingfisher data were used to test the three general hypotheses seeking to explain proximate causes of species-diversity gradients in terms of niche metrics: the niche-breadth, niche-overlap, and resource-base mechanisms. The data support the resource-base hypothesis and, to a lesser extent, the niche-overlap hypothesis. That niche-overlap values were somewhat more constant than expected by chance is interpreted as support for the concepts of diffuse competition and limiting similarity. However, the number of uncontrollable variables and problems with interpretation of indices of community structure makes rigorous testing of hypotheses impossible. Current hypotheses concerning the ultimate biotic factors producing species-diversity gradients were also examined. Although the kingfisher data provide only weak evidence for or against most hypotheses concerning tropical species diversity, one hypothesis can be strongly rejected: diversification with time. Piscivorous kingfishers in the New World reach single-point species richness values (five syntopic species) equivalent to or greater than those in the Old World. Because New World kingfishers are almost certainly more recently evolved relative to Old World kingfishers, relative age of kingfisher communities cannot be invoked to explain differences in single-point diversity.

INTRODUCTION

One of the central questions of community ecology is "What determines the number of species that can coexist in community?" This question has been asked most frequently in the context of latitudinal gradients in species diversity. For centuries biologists have puzzled over the much greater diversity of organisms in the tropics than in temperate regions, and the impoverished biota of the poles. G. E. Hutchinson and Robert MacArthur focused community ecology on the problem of tropical species diversity, generating a substantial body of primarily theoretical literature in the 1960's that dealt with the question "Why do some communities have more species than others?" and especially, "Why do the tropics have so many more kinds of organisms?" Unfortunately, even at present, few quantitative field studies exist that test the hypotheses concerning latitudinal gradients in species diversity. The purpose of this research was to provide one such study, one that focuses on diversity patterns in New World Kingfishers (Alcedinidae).

The ultimate cause of tropical species diversity is, of course, the equatorial position of the tropics on a tilted globe. This may seem trivial and obvious, but it is seldom mentioned in discussions of species diversity. Only MacArthur (1972, Ch. 1) properly emphasizes this and its consequences in terms of climatic patterns at the equator: energy received from the sun is increased, seasonality is reduced (because the earth's axis is tilted), and mean annual precipitation is increased (because of trade-wind patterns set up at the equator). These physical factors are the foundation for a second level of ultimate causes, the ecological factors, such as resource specialization, interspecific competition, and predation, that transfer the effects of these physical factors to organisms and communities. Field experiments that measure these ecological factors are usually difficult to execute, particularly for vertebrates in the tropics. For instance, measuring predation rates or intensity of competition, much less manipulation of these parameters, requires systems with special features that permit such measurements.

One view of the species diversity question holds that species richness is simply the product of the balance between speciation and extinction events and that geographic differences in species richness reflect historical differences in this balance; interspecific interactions, particularly interspecific competition, plays little if any role in determining this balance. Although the importance of historical factors cannot be disputed in many cases, I feel that their role is negligible in determining current patterns of species richness of New World kingfishers for two reasons. First, at present there are virtually no barriers to dispersal of kingfishers from Canada east of the Rockies south to central Argentina east of the Andes. Unless one proposes that the current distributions are not in equilibrium (a proposal not supported by any known, ongoing changes in kingfisher distribution), the current distribution patterns must be the result of current ecological conditions. Second, although kingfishers have been present in the New World for much less time than in the Old World (see Discussion), single-point species richness in kingfishers in the New World is the same as in the Old World, contrary to the predictions of a history-based hypothesis. Therefore, this project was conducted under the premise that current ecological conditions produce the geographic patterns of species richness.

A second view of the species diversity question, one that is an ecologically oriented derivative of the historical hypothesis, holds that current patterns are the result of patterns of resource availability and single-species responses to these patterns, and that interspecific competition again plays little if any role. The fieldwork for this study was conducted from 1974 to 1977, at the twilight of a period when the importance of interspecific competition in determining diversity patterns was not questioned (Schoener 1982, Wiens 1984). Thus, the framework of the study and many of the initial interpretations (Remsen 1978) reveal the prevailing, competition-dominated viewpoint of that time. The best way to distinguish between the competition hypothesis and autecology hypothesis is by experimental manipulations of the resource base and of the potentially competing species; such manipulations were not feasible in this study. Even with the benefit of hindsight, not much data can be mustered to distinguish between the competition and autecology hypotheses. All that can be said is that the five species of Neotropical kingfishers investigated herein represent a system in which the effects of ongoing interspecific competition are "likely" for the following reasons: (1) overlap in prey size selection, feeding site selection, and habitat use is high (see appropriate sections below), (2) distribution of body sizes within the five species conforms to the predictions of competition-induced character displacement (Remsen, in prep.), and (3) ecological parameters such as prey-size and perch-height selection are flexible rather than fixed.

The proximate causes of tropical species diversity are much more amenable to study. Three hypotheses have been put forward to explain how species are packed into, or more correctly, unpacked from tropical communities: the niche-breadth mechanism, the niche-overlap mechanism, and the resource-base mechanism (see Discussion). If one accepts the niche concept and the structure imposed upon communities by Hutchinson's (1957, 1965) definition of the niche, then these three mechanisms, although not mutually exclusive, are the only ways that species can be packed into communities under the constraints of niche metrics in one dimension, regardless of whether these are induced by interspecific competition or single-species responses to resource availability. These three hypotheses can be tested in the field. The outcome of these tests can provide evidence for which of the ultimate factors play the most prominent roles in influencing species diversity gradients.

These proximate mechanisms will be discussed more thoroughly after presentation of the core natural-history data of the kingfisher system investigated.

The purpose of this research was to test these three hypotheses concerning the proximate causes of species-packing. Rather than attempt to do this for an entire community of organisms or even a major subset of the community, such as fish-eating vertebrates, the design of my study was to focus on a small group of taxonomically and ecologically similar organisms, the five species of Neotropical kingfishers.

The kingfishers were chosen for several reasons. First, the latitudinal gradient in kingfisher diversity parallels that for most taxonomic groups: the center of species richness is in the equatorial regions, with five species, and this declines in an orderly manner in both directions toward the poles, until only one species is found in the temperate regions of both hemispheres (Figs. 1, 2). Second, all New World kingfishers feed in a similar way on similar foods, maximizing chances for interspecific interactions and parallel responses to changes in resource availability. Third, the physiognomy of kingfisher habitat changes little along the latitudinal gradient: the basic structure of kingfisher habitat, i.e., water bordered by trees, remains essentially unchanged. Kingfishers are negligibly affected by radical changes in plant species composition or, more important, by the horizontal and vertical habitat structure and heterogeneity and floristic composition that are thought to influence terrestrial communities so strongly (e.g., MacArthur et al., 1962; MacArthur 1964; Karr and Roth 1971, Johnson 1975, Holmes et al. 1979, Holmes and Robinson 1984, Robinson and Holmes 1984, Holmes and Recher 1986). Thus the importance of latitudinal differences in within- and between-habitat heterogeneity or floristic composition is minimized. Fourth, kingfisher feeding behavior has several features that make these birds attractive as study organisms. Prey items are large relative to body size and thus not immediately swallowed; after capture, a prey item is beaten to death and manipulated into a head-first position before it is swallowed. This allows an observer to estimate the size of the prey item and, in some cases, to identify it with varying degrees of taxonomic precision. Also, the precise location at which prey is captured can be pinpointed: concentric ripples from the dive into the water are often visible for nearly a minute after entry. MacArthur (1972: 133) has previously pointed out the interesting possibilities posed by the kingfisher system.

The following questions were asked for this kingfisher system: What determines the number of species that can coexist in a given area? Why do kingfishers display a latitudinal gradient in species diversity? And what is the proximate mechanism involved in changes in species-packing in kingfishers?

Ideally, study sites with all possible combinations of kingfisher species (5, 4, 3, 2, and 1 species) would have been examined and replicated in both directions from the equator and at the same sites in different years. It was obvious, however, after one field season at a five-species site that not enough time was available to obtain sufficient sample sizes for all species in all five combinations. In fact, it was a major effort just to obtain sufficient sample sizes in the critical, maximum species-packing, five-species situation in a single year, and quantitative data were obtained for only one other situation, the three-species system. Thus the basis of this study is the comparison of the five-species system with the three-species system, and the possibility of between-year differences had to be ignored. The five-species system was examined at two sites: one in the geographic center of the five-species system and one at the periphery. The rationale was that it seemed necessary to see whether niche metrics changed significantly between center and periphery, before examining other combinations of species.

The advantages of studying kingfishers were somewhat counterbalanced by some difficulties. The largest species, *Ceryle torquata*, was wary and difficult to approach. Furthermore, individuals were often followed for two hours or more without recording a single successful capture. Thus a large time investment was required to obtain sufficient sample sizes for prey items. The smallest species, *Chloroceryle aenea*, also posed problems. It favored overgrown, densely vegetated portions of stream and lake shorelines, where observation was difficult. As found by Willard (1985), it dived frequently and changed location constantly and rapidly between dives, making it difficult to follow. It was also the least common of the five species; as summarized by Forshaw (1983), most authors have described this species' status as "rare" in most of its range. A third species, *C. inda*, also presented observational problems, because it foraged along the same overgrown shorelines as *aenea*. Only *C. americana* and *C. amazona* can be recommended as organisms suitable for studies needing large sample sizes for critical data such as prey items (e.g., see Betts and Betts 1977).

NEW WORLD KINGFISHER DISTRIBUTION

Six species of kingfishers (Coraciiformes: Alcedinidae) in two sister genera inhabit the New World: *Ceryle torquata* (Ringed Kingfisher), *Ceryle alcyon* (Belted Kingfisher), *Chloroceryle amazona* (Amazon Kingfisher), *Chloroceryle americana* (Green Kingfisher), *Chloroceryle inda* (Green-and-rufous Kingfisher), and *Chloroceryle aenea* (American Pygmy Kingfisher; or, to avoid confusion with Old World species with a similar name, Least Kingfisher, as proposed by Blake [1953]).

The maximum species density, five sympatric species, characterizes much of the lowland tropics (Figs. 1 and 2; see Remsen 1978 for details): most of the Amazon and Orinoco basins, the Guianas, lowland southeastern Brazil, the Caribbean and Pacific lowlands of Colombia, the Darién region of Panama, and a region extending north in Panama in the Caribbean lowlands. Although five species have been recorded in Caribbean Costa Rica and Nicaragua, *inda* evidently occurs only very locally in these areas (AOU 1983). The five-species assemblage corresponds well with the distribution of the main mass of humid, lowland tropical forest, although this habitat extends farther north in the Northern Hemisphere and perhaps farther south in South America than does the five-species kingfisher system.

Proceeding away from the equator in either direction, species begin to drop out. First to do so in the north is *inda*. This may also be the case in South America, but data are lacking from critical areas. The next species to drop out in the north is *aenea*, but in South America it is again unknown if *aenea* extends farther south than does *inda*. The third species to leave the system in both north and south is *amazona*. Three areas are seem to be

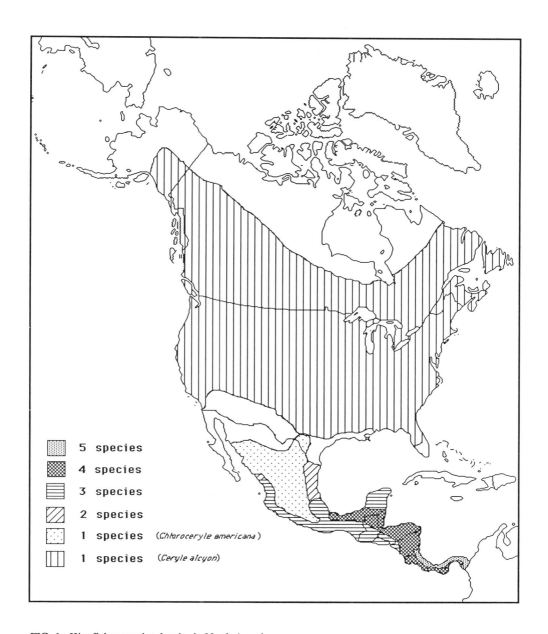

FIG. 1. Kingfisher species density in North America.

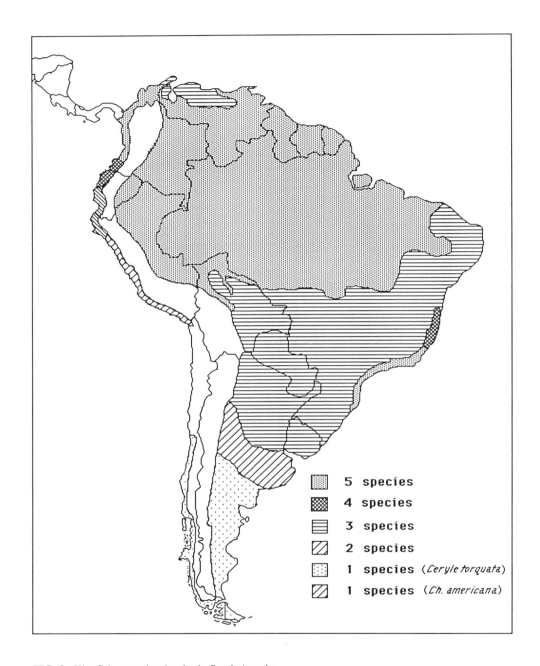

FIG. 2. Kingfisher species density in South America.

exceptions: *amazona* is absent from western Ecuador, Trinidad, and evidently Yucatan, whereas *aenea* is present in all three regions and *inda* is present in western Ecuador.

The fourth species to drop out in North America is *torquata*, leaving *americana* as the sole species in most of arid interior and extreme northern Mexico, or *alcyon* in most of the United States and Canada. These latter two species are apparently sympatric in a small area in the Edwards Plateau in Texas (Oberholser and Kincaid 1974). In southern South America, where *alcyon* does not occur, *torquata* extends through latitudes occupied in North America by *alcyon* (as pointed out by MacArthur [1972: 133]). Because *torquata* extends no farther north in North America than the southern limit of the breeding distribution of *alcyon*, and because the climate in southern South America is as harsh as that of at least the southern third of the breeding range of *alcyon*, the presence of *alcyon* in North America probably limits the northward expansion of *torquata*, even though these two species are not precisely parapatric at present.

A distinctive subspecies, *C. torquata stellata*, is found mainly in that area of the distribution of *torquata* where it is the only kingfisher species. This subspecies differs from nominate *torquata* in having a shorter, more *alcyon*-like bill length (Table 1): the mean culmen length of *C. t. stellata* is 63.5 mm, over 11 mm shorter than that of *C. t. torquata*. The dorsal coloration of *stellata* is also a darker blue than that of nominate *torquata*, tending toward that of *alcyon*. These trends suggest the possibility of parallel evolution between *stellata* and *alcyon*.

Reflecting the milder climate of southern South America, the latitude at which species numbers change is 10-15° farther north than the comparable change-over latitudes in North

TABLE 1. Bill lengths of adult kingfishers. Measurements of the first five taxa are taken only from areas where all five are sympatric.

Species	Bill length (exposed culmen)			
	Mean (mm)	S.D.	Range (mm)	N
Ceryle t. torquata	75.0	3.70	66.2 - 85.8	87
Chloroceryle amazona	64.0	3.50	54.5 - 72.9	161
Chloroceryle inda	47.0	2.28	43.9 - 52.2	120
Chloroceryle americana	39.3	2.11	33.0 - 44.1	100
Chloroceryle aenea	27.6	1.37	24.9 - 31.0	39
Ceryle alcyon	52.6	2.95	46.3 - 61.6	127
C. t. stellata	63.5	3.40	57.2 - 69.2	26

America. The "poleward" limit of the two-species system is at about 27° N in North America, but about 41° S in South America. The limit of the three-species system in North America is at about 23° N, compared to about 34° S in South America. For the five-species system, the limit in North America is at about 10-13° N, compared to about 27° S in South America.

Topographic and climatic variations also modify the effects of latitude on species diversity. Areas without humid tropical forest, such as western Peru, northern Venezuela, and the Pacific coast of much of central America, have fewer species than would be predicted from their latitudes. Kingfisher species richness decreases with increasing elevation; this is almost certainly because fish populations decrease as, with increasing elevation, current strength and turbulence of rivers and streams increase and productivity decreases. The fish fauna of Amazonia declines rapidly above 300 m, with a 70-75% reduction in species richness even as low as 600 m (Lowe-McConnell 1975). Even if fish were abundant in the cascading torrents of the Andes, capturing them would be extremely difficult. The sequence in which kingfisher species leave the system with increasing elevation on a single elevational gradient is largely unknown. The species most restricted to the lowlands are *inda* and *aenea*, and the species found at highest elevations tend to be *torquata* and *americana* (Hilty and Brown 1986), paralleling the latitudinal pattern. Examination of nearly 1000 specimens from South America revealed only 42 taken above 500 m, distributed as follows: (a) *americana*, N = 24, maximum elevation 2684 m; (b) *amazona*, N = 8, maximum elevation 1068 m; (c) *torquata*, N = 6, maximum elevation 1068 m; (d) *aenea*, N = 1, 700 m; and (e) *inda*, N = 1, 600 m. Therefore, the elevational pattern resembles the latitudinal pattern in the sequence with which species drop out; because *torquata* is more difficult to collect than *amazona*, I suspect that use of specimen records is biased against *torquata* and that *torquata* actually occurs higher and in greater numbers than *amazona*. The only species regularly found above 1100 m is *americana* (N = 11).

Species richness of kingfishers is also reduced in areas of savanna within the latitudes that generally have five species. For example, a locality in the "llanos" of Venezuela lacks *inda* (Thomas 1979), and both *inda* and *aenea* occurred only as wanderers at my study site in the Bolivian "pampas" (Remsen 1986; see Study Sites section).

The regions without breeding kingfishers are generally those areas too high in latitude or elevation to have trees (for hunting perches), so steep that rivers are too turbulent for kingfisher foraging, or so dry that suitable streams are virtually nonexistent, such as much of the southwestern United States, extreme northern Mexico, Baja California, and northern Chile.

Some areas seem to have suitable kingfisher habitat but lack breeding kingfishers. The lower Colorado River in the southwestern United States is such an enigma. Trees for hunting perches and banks suitable for nest tunnels are all present, but *alcyon* is present only during the non-breeding season (Rosenberg et al., in press); perhaps dramatic water-level fluctuations and silt-laden, turbulent waters of such rivers of the arid southwestern United States reduce populations of surface fishes to levels that cannot support kingfishers (Grinnell 1914). Even more perplexing is the absence of breeding kingfishers from most of the West Indies, especially Cuba, Hispaniola, Puerto Rico, and Jamaica, where large streams are numerous. All habitat requirements seem to be met, *alcyon* winters extensively in the West Indies (AOU 1983), and *torquata* is present on three oceanic islands in the Lesser Antilles (AOU 1983). Although the absence of the other four New World species could be explained by poor transoceanic dispersal abilities, why these islands have not been colonized by *alcyon* or *torquata* is a mystery.

STUDY SITES

Three study sites were chosen, one in Colombia and two in Bolivia. The Colombian site, about 40 km northwest of Leticia, Amazonas, Colombia, consisted of two subsites: an island, Isla de Santa Sofía II (hereafter ISS), in the Amazon River, and a stream, Quebrada Tucuchira, on the Colombian mainland near the island. ISS, about 4 km long and 1-1.2 km wide, contained several large lakes connected to the Amazon River during the March to June flood-season. These lakes were bordered by forest, but a mat of floating aquatic vegetation (mainly *Paspalum* and water hyacinth) of varying width fringing the shoreline reduced the availability of suitable kingfisher habitat by increasing the distance from hunting perches to the nearest open water. Thus, although the shoreline was lined with suitable elevated hunting perches, many of these overlooked dense mats of aquatic vegetation rather than open water suitable for diving. In contrast, the stream site had only small patches of floating vegetation, and with forest lining its banks, suitable hunting perches were distributed continuously along the shore. Most forest along the stream banks was at least seasonally, if not permanently, flooded. Several smaller streams flowed into the main stream. These tributaries were so small (1-3 m wide) that the forest canopy was continuous overhead; in contrast, the main stream was sufficiently wide (10-20 m) that the canopies of the forest on opposite banks did not touch.

The first Bolivian study site was Lake Tumi Chucua, a large oxbow lake at Tumi Chucua, near Riberalta, Beni, Bolivia (see Pearson [1975] for description of area and its avifauna). This lake connected with a major Amazon tributary, the Río Beni, during flood season. Forest was nearly continuous along the shoreline, with unlimited availability of hunting perches except at the north end of the lake, where there were extensive mats of floating aquatic vegetation. Thus, the shoreline was more similar to that of the Colombian stream site, Tucuchira, than to that of ISS. The similarity was further increased by the presence of extensive areas of partly flooded shoreline. Although humid lowland tropical forest surrounds Tumi Chucua in all directions, this was near the southern limit of the continuous forest of the Amazon Basin. Just 50 km south of Tumi Chucua appear the first patches of grasslands, or "pampas," that mark the beginning of the savannas characteristic of north-central Bolivia.

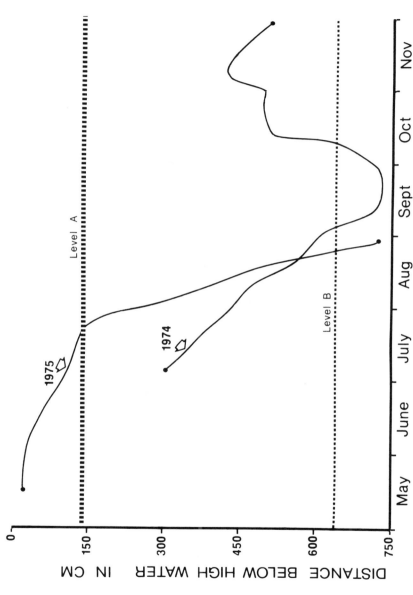

FIG. 3. Water-level fluctuations of the Amazon River at Isla de Santa Sofía, Amazonas, Colombia. Daily measurements of distance below high-water mark on river-bank trees were plotted for the two periods of residence in 1974 and 1975. Level A indicates that river level at which the lakes on Isla de Santa Sofía became disconnected from the river, the point at which silt-laden river water was no longer present in Quebrada Tucuchira, and the point at which substantial areas of land on the island and nearby mainland became exposed by the receding river. Because of these significant ecological changes, this level was used to define the boundary between high-water and low-water season. Apparently, 1975 was a rather late year: the river remained high for two to three weeks longer than normal. Level B was the river level at which mainland streams such as Quebrada Tucuchira flowed steeply downhill to the river.

The second Bolivian site was about 250 km south of Tumi Chucua, in the savanna country, at Estancia Inglaterra, a cattle ranch along the Río Yata, a small tributary of the Río Beni. Here, forest was restricted to the shorelines of rivers and streams and also to certain patches of high ground. The principal habitat was tall grasslands interspersed with savanna (see Remsen [1986] for more detailed description of the site and its avifauna). Most data were gathered along Arroyo Salsipuedes, a small stream flowing into the Río Yata. This stream was only 4-8 m wide, much smaller than Tucuchira, and the trees bordering it were also much smaller than those at Tucuchira, so the canopy did not close over the stream. Although Arroyo Salsipuedes was a smaller and shallower stream than Quebrada Tucuchira, this evidently did not limit the size of aquatic organisms inhabiting it. In addition to many fish much larger than the largest taken by kingfishers, 2-3 m long individuals of the Amazon dolphin *Inia geoffrensis,* and the spectacled cayman *Caiman crocodilus*, were seen in the stream.

All these sites and their various subsites are described in greater detail below in relation to habitat differences in kingfisher density.

Water levels showed marked seasonal fluctuation. Those for ISS and the adjacent Amazon River are plotted in Figure 3. No data are available for Lake Tumi Chucua, but the lake probably follows a similar pattern. Data are likewise lacking for Arroyo Salsipuedes, but from information from local residents, it was evident that this stream showed the greatest water-level fluctuations of any of the study sites. It reportedly became completely dry during parts of the dry season, as did occasionally much of the Río Yata itself except for standing pools left in deeper channels and oxbows. In the flood season the stream overflowed its banks, and the area for miles around was mostly under water.

ISS and Tucuchira were visited in August 1973, from July to November 1974, and from May to August 1975. Tumi Chucua was visited in November 1976 and January 1977. Arroyo Salsipuedes was visited in November and December 1976. I was present during the extremes of seasonal water fluctuation at each site, minimizing the chances of missing some important phase in kingfisher feeding ecology. I also observed little change in feeding behavior within either of the two seasons, high-water and low-water, and this provides some reassurance that some critical phase of kingfisher ecology has not been missed. Nevertheless, the possibility exists that year-round data would reveal different patterns, and the question of annual variation cannot be addressed at all.

FISH DATA

Kingfishers rely almost completely on surface fishes for food. Therefore, I invested much effort in quantifying densities of surface fishes at the intensive study sites to give an index of the size and structure of the resource base at each site.

Several methods for sampling fish populations to compare prey availability between sites were attempted, including gill-netting, cast-netting, and poisoning, but each had its difficulties in implementation and interpretation of results. The method decided upon — visual counting of surface fishes — not only was the most readily adaptable to all field situations, but also seemed to approximate most closely the way that kingfishers view their prey, because kingfishers seldom took fishes below the first centimeters of the surface. Even the largest kingfisher species seldom disappeared underwater upon impact, despite considerable momentum produced by the dive. The probability of successful capture may decline by some power function with the target's distance below surface for several reasons: (1) the deeper the target, the greater the distortion of its true location, because of refraction; (2) the characteristically silt-laden, murky water drastically reduced the visibility of prey items below the surface — at some localities, fish were visible only within a few centimeters of the surface; and (3) the deeper the fish, the longer the time between the kingfisher's impact with the surface and contact with the target prey, and this interval probably increases exponentially with distance from surface because of deceleration of the kingfisher through the water (even without braking action by the kingfisher). Only once did a kingfisher disappear completely below the surface for an interval that indicated that prey was captured at some depth below the surface: an *amazona* was once seen to capture a loricariid catfish (probably *Hypostomus* sp.) about 30 cm below surface. (These catfishes were regularly seen gleaning from submerged branches.)

SAMPLING METHODS

The following procedure was developed for sampling surface fishes. A semicircle with a 2-m radius was watched for 20 minutes from a canoe, during which time every fish visible within the semicircle was counted and its size estimated to the nearest inch. (Here and throughout the study, I have retained nonmetric units in the analyses for data based on visual estimates using inches and feet; although the nonmetric units may be irksome to some, retention of nonmetric units was preferable to conversion to awkward or nonsensical metric intervals, e.g., the "2.2-4.4 cm" size category vs. "1-2 inch" category). Each fish was identified as precisely as possible, almost always to family, sometimes to genus, and occasionally to species. Voucher specimens of most, common surface fishes were deposited at the California Academy of Sciences, San Francisco. Only fish within 20 cm of the surface were counted. Each fish was counted only once, even if it swam out of the sample area and returned (although some duplication undoubtedly occurred, because some delayed returners could not be distinguished from newcomers of the same size and species). Only fish present at least one second were counted, because fishes visible for shorter periods were presumably unavailable to kingfishers.

The usual procedure was to ease the canoe slowly to the selected sampling point on the shoreline and to lodge it slightly in vegetation to prevent further movement. A 2-m radius semicircle was visualized, using measure-marks on the canoe to delimit the 4-m diameter. After five minutes had elapsed, to allow local fishes to adjust to the presence of the canoe, the count was begun. Trials indicated no significant difference between fish counts in the first and last five minutes of the sample period; thus, I assumed that the presence of the canoe did not affect the presence of fishes. Fish in size categories preyed upon by kingfishers ignored the canoe as long as I made no rapid movements. In addition to fishes, shrimp and crabs were counted because these were also captured occasionally by kingfishers.

Each study site was sampled systematically to include almost all the shoreline and all hours of the day. Fish were easier to see in sunlit or clear water, but presumably kingfishers find them easier to see under these conditions, too. Most effort was concentrated on the first 2 m from effective shore, i.e., one sample-area radius from shore, because 85-100% of kingfisher dives were in this zone (see Dive Entry section). All surface fish data presented below refer to the first 2 m from effective shore, unless stated otherwise. This sampling technique approximates as closely as practicable the way in which kingfishers view prey densities. Ideally, the observer should be on a lofty perch for a more kingfisher-like vantage point, rather than sitting in a canoe.

Fish density at Lake Tumi Chucua was sampled only to the extent that it could be shown to be statistically indistinguishable from fish densities in similar habitats at Quebrada Tucuchira and ISS. Thus, fish-density data from "five-species habitats" contain little data from Lake Tumi Chucua. The fish fauna of Lake Tumi Chucua has been studied by Swing and Ramsey (1984, 1987).

FISH DENSITIES

Among the study sites during low-water season, the lakes on ISS showed the highest surface densities in most size-classes of fishes, followed closely by the stream study-sites at Tucuchira (Table 2). Shaded habitats had only 25-30% of the density of the other two, although the differences were mainly in fishes under 3 in. in length; larger fishes showed equivalent or higher densities in shaded habitat.

TABLE 2. Surface fish densities (mean number individuals/census) at the intensive sites.

							Fish Size-class					
Site	Habitat	Season	N	All	1 in.	1-2 in.	2-3 in.	3-4 in.	4-5 in.	5-6 in.	6-7 in.	>7 in.
ISS	lakes	low-water	54	85.6	27.1	39.0	15.7	2.4	0.7	0.3	0.2	0.2
Tucuchira	streams	low-water	46	62.9	12.8	36.9	11.9	0.9	0.3	<0.1	<0.1	—
Tucuchira	shaded	low-water	54	20.0	8.3	7.2	1.3	0.9	1.2	1.0	<0.1	<0.1
ISS	lakes	high-water	35	3.7	2.8	0.5	0.3	<0.1	<0.1	<0.1	<0.1	—
Tucuchira	streams	high-water	38	2.4	0.3	1.3	0.5	0.1	0.1	0.1	<0.1	—
Tucuchira	shaded	high-water	46	7.6	5.8	1.6	0.1	<0.1	<0.1	<0.1	<0.1	—
Salsipuedes	streams	low-water	46	1.2	0.5	0.4	0.1	0.1	0.1	<0.1	—	—

During high-water season, densities in all habitats dropped dramatically, especially in stream and lake habitats, where high-water densities were only 3.8% and 4.2%, respectively, of low-water densities (Table 2). Seasonal decrease in density in shaded habitat was much less: high-water density was 37.8% of low-water density. Consequently, shaded habitat showed higher surface-fish densities than either lake or stream habitats during high-water season, reversing the trend at low-water season. During high-water, densities on the lake and stream sites were only 47.4% and 31.6%, respectively, of shaded-habitat densities.

Densities at Arroyo Salsipuedes, the study site with only three species of kingfishers, were lower than those of any of the habitats at any season in the five-species study sites (Table 2). Densities of fishes larger than 3 in. or those smaller than 1 in., however, did not differ from those in study sites with five kingfisher species at high-water season.

The patterns of seasonal change among fish size-classes differed dramatically in the three habitats (Fig. 4). At the stream site of Tucuchira, the difference between high- and low-water surface densities increased exponentially from large to small fish, whereas it decreased exponentially in shaded habitat at the same locality. Thus, seasonal differences in densities at the stream site were due mainly to the decline in density of small fishes; the larger the fish, the less the proportional change in density. The reverse was the case for shaded habitats, where surface fish densities showed the smallest seasonal differences for small fishes and the greatest for large fishes. In the lakes, a third pattern was evident: decreases in density at high-water were more or less uniform among size-classes of fishes.

Why does fish density differ among the habitats? First, on a geographic level, the site with three kingfisher species, Arroyo Salsipuedes, is almost certainly a more strongly seasonal and more unpredictable environment. According to local residents, most of Arroyo Salsipuedes dries up completely during the dry season, as does much of the adjacent Río Yata. Water is left standing in the deeper channels and in miniature oxbow lakes, but the main flow of surface water ceases. Such dramatic dry-season effects are evidently typical of the savannas of eastern Bolivia (and elsewhere in the tropics [Lowe-McConnell 1975]). Environmental fluctuations of this magnitude presumably depress fish populations. In contrast, none of the other intensive sites showed this degree of seasonality. Although water levels fluctuated as much as 9 m in the main channel of the Amazon River, fluctuations on the lakes of ISS and the stream at Tucuchira were much less (Fig. 3). When both the lakes and the stream were at the same level as the Amazon River during flood season, their water-level fluctuations paralleled those of the main river. But the lakes were connected to the river only during the high-water season, March-June; and after becoming disconnected when the river level dropped, water levels in the lake remained relatively constant for the rest of the year, changing only 1-2 m in response to seasonal rainfall trends. Quebrada Tucuchira remained at the same level as the river for a much longer period. The only difference was that during the peak of the low-water season from late August to early October, when the Amazon experienced its most severe drop in water level, Tucuchira dropped only slightly; consequently, water flowed strongly for the last several hundred meters (section A) of the stream. But only during high-water season did the murky, silt-laden water of the Amazon penetrate Tucuchira; the remainder of the year, the water was

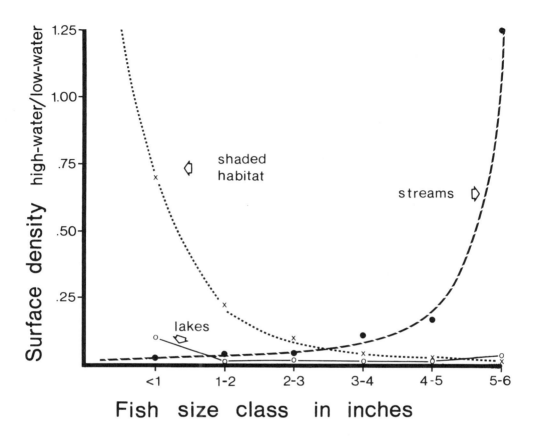

FIG. 4. Patterns of seasonal change in fish density in different size classes in three habitats. The ratio of high-water to low-water density was used as the measure of seasonal change.

much clearer, with a sharp boundary between clear and murky water at the boundary between sections A (the first 100 m) and B (the next 100 m upstream).

On a day-to-day basis, water-level fluctuations at Salsipuedes were much greater than at Tucuchira or ISS. After a heavy rain at Salsipuedes, water rose as much as 2 m overnight. In contrast, the maximum single-day rises after rain at Tucuchira and ISS were under 15 cm.

Only speculation is possible concerning the differences in surface fish densities among the three major habitat types (lakes, streams, and shaded habitat). At low-water season, the greatest concentrations of surface fishes occurred where the floating vegetation mat was most extensive. Within a given habitat type, sample points with floating vegetation, particularly *Paspalum*, had the greatest surface fish counts. These mats, or "floating meadows," support large concentrations of invertebrates that presumably form the food base for the surface fish fauna (Junk 1970, 1973). In shaded habitat these vegetation mats were absent, and surface fish density was much lower than in either lakes or in streams where floating vegetation mats were extensive. Lakes, with the highest surface fish densities of any habitat, were almost completely lined with floating vegetation mats. At stream sites, those sections with the greatest development of these mats had the highest surface fish counts.

Surface fish densities of two sites were not included in Table 2 because they could not be considered part of any of the three major habitat types: a section of the Amazon River adjacent to ISS and the last 200 m (Section A) of Quebrada Tucuchira (see next chapter for descriptions). Both these sites had extremely low surface fish densities despite the presence of floating vegetation mats, extensive at Tucuchira A and sparse on the Amazon River. Water clarity at both these sites was extremely low; the water was so silt-laden that fish could not be seen deeper than 2 cm. Thus, in addition to whatever inhibiting effect extremely murky water might have on surface fish populations, there definitely was a negative effect on ability to census fish, even if they were present, and this reduced detectability presumably affected kingfishers as well as the fish-counting observer. The strong current in the river channel may also have had a depressing effect on populations of very small fishes, which would presumably be less able to cope with such a current than could larger fishes.

Why does fish density decline so drastically at high-water? It was assumed that this was because the fish population spreads out in response to the extensive flooding of areas adjacent to the Amazon, greatly expanding the amount of available habitat. The terrain adjacent to the river is flooded from 200 m to several km inland from the banks, depending on local topography. High-water season seems to be the breeding season for many of the Amazon Basin fishes, which capitalize on the newly flooded areas, which are rich in food (Roberts 1973, Lowe-McConnell 1975, Goulding 1980). Such areas experience only a weak current, and silt consequently settles here more rapidly than elsewhere, improving water clarity and, presumably, the suitability of these areas to fishes. The tremendous decline in water clarity on the lakes and streams as they become inundated by water from the Amazon reduces fish detectability and further accentuates the decline in surface fishes.

The reasons for the differences in the pattern of seasonal change for the various size-classes (Fig. 4) are unknown. Only a thorough study of seasonal movements and ecology of the local fish fauna would provide plausible explanations. Kushlan (1976a) found that during an unusual period of high-water, mean body size of fishes increased because of an influx of larger, piscivorous fish species; unfortunately, my data on surface fishes could not be analyzed with respect to trophic level.

SIZE-CLASS PROPORTIONS

Comparisons of proportions of the surface fish population in various size classes among the three habitats during the low-water season (Table 3) yielded no significant differences (Kolmogorov-Smirnov test, $P > .05$) in distribution of proportions, although the

TABLE 3. Proportions of surface fish population by size class.

							Fish Size-class				
Site	Habitat	Season	N	1 in.	1-2 in.	2-3 in.	3-4 in.	4-5 in.	5-6 in.	6-7 in.	>7 in.
ISS	lakes	low-water	54	.312	.457	.184	.028	.008	.004	.002	.001
Tucuchira	streams	low-water	46	.204	.587	.190	.015	.005	.001	.001	—
Tucuchira	shaded	low-water	54	.414	.359	.064	.046	.061	.048	.002	.008
ISS	lakes	high-water	35	.776	.134	.078	.008	.003	.003	.001	—
Tucuchira	streams	high-water	38	.133	.556	.222	.044	.022	.022	.001	—
Tucuchira	shaded	high-water	46	.760	.209	.017	.006	.005	.002	.001	—
Salsipuedes	streams	low-water	46	.435	.391	.086	.044	.001	.001	.001	—

completely shaded habitat was close to being significantly different from streams (P < .10) in having a greater proportion of very small and very large fishes.

When comparisons were made between seasons within the same habitat, fish size distributions in streams did not change significantly between high- and low-water. In lakes and shaded habitats, however, seasonal changes were significant (P < .005): tiny fishes became disproportionately common at high-water. When comparisons were made between Arroyo Salsipuedes (the three-kingfisher-species study site) and the stream habitat at Tucuchira (the five-species study site), Salsipuedes was found to have a significantly (P < .05) lower proportion of large fishes. The size distribution of surface fishes at Salsipuedes most closely resembled that of the completely shaded habitat at low-water.

A substantial majority (at least 83% in all samples) of all surface fishes were under 3 in. (7.6 cm) in length. In general, the larger the fish, the rarer it was on the surface, except that the smallest size-class, less than 1 in., did not always form the greatest proportion of fishes. In three of the habitat-season combinations — streams at both seasons and lakes at low water — the next-largest category, 1-2 in., was the dominant size-class, and 2-3 in. fish were nearly as common or more common than those smaller than 1 in. (Table 3). In two other combinations, shaded habitat at low-water and Salsipuedes at low-water, 1-2-in. fish were only slightly less numerous than those under 1 in.

Thus, in only two season-habitat combinations, high-water in lakes and in shaded habitat, did the smallest fish size-class form a substantial proportion of the surface fish population (Table 3). Furthermore, fishes under 0.5 in. (1.1 cm) were extremely rare on the surface; although the proportion of individuals in this category was not quantified, it was qualitatively obvious that this size-category was virtually nonexistent on the surface of open water. Yet if a total fish community census could be taken, certainly the smallest fishes would be the most numerous, at least during those months following the breeding season. Apparently the bulk of the tiniest fishes and fish fry are found away from the surfaces — or if on the surface, where covered with protective floating vegetation.

The preceding data on size-distribution patterns applied only to the first 2 m from "effective" shore. The proportion of fishes in larger size-categories increased greatly with increasing distance from shore. Most censuses taken farther than 2 m from shore recorded no fishes at all, partly because fish density declined rapidly with increasing distance from shore, and partly because large fishes were more wary.

TAXONOMIC COMPOSITION

The majority of surface fishes were members of the family Characidae, a diverse group that reaches peak diversity in the Amazon Basin. From 75% to 95% of all surface fishes visually censused within 2 m of shore were identified as characins (Table 4). From a sample of about 150 voucher specimens from the Colombian sites, only 38 could be identified to genus without further aid of experts on the genera involved. The identified genera were, in decreasing order of numerical abundance in the sample, *Moenkhausia, Hyphessobrycon, Astyanax, Aphyocharax, Roeboides,* and *Curimata.* Species in three of these genera

TABLE 4. Proportions of Characidae in surface fish populations. The proportion is calculated as the number of individual Characidae divided by the total number of individual fishes counted in that size class. Blank cells indicate that no fishes of any kind were counted in that particular sample.

					Fish Size-class						
Site	Habitat	Season	N	All	1 in.	1-2 in.	2-3 in.	3-4 in.	4-5 in.	5-6 in.	6-7 in.	>7 in.
ISS	lakes	low-water	54	.767	.726	.835	.728	.313	.271	.889	.600	.000
Tucuchira	streams	low-water	46	.902	.956	.904	.879	.476	.000	.000	.000	—
Tucuchira	shaded	low-water	54	.750	.929	.564	.457	.920	.758	.961	.000	.000
ISS	lakes	high-water	35	.950	.979	.823	1.000	1.000	.000	.000	.000	—
Tucuchira	streams	high-water	38	.847	.970	.946	.920	.900	.500	.250	.000	—
Tucuchira	shaded	high-water	46	.867	.967	.920	1.000	.500	.000	.000	.000	—
Salsipuedes	streams	low-water	46	.837	.960	.768	.901	.800	.500	.000	.000	—

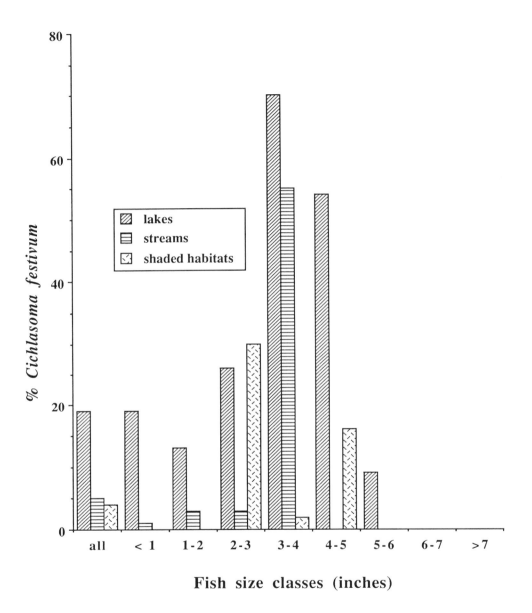

FIG. 5. Proportion of *Cichlasoma festivum* in surface fish censuses in various size classes for three habitats.

(*Moenkhausia, Hyphessobrycon,* and *Roeboides*) were among the most common fishes at Tumi Chucua (Swing and Ramsey 1987).

In the visual censuses of surface fishes, one species of cichlid, *Cichlasoma festivum,* formed the bulk of the non-Characidae component at low-water season, particularly in the lakes at ISS (Fig. 5). Of the remaining non-characins, the following were recorded, listed here in decreasing frequency of occurrence: *Rivulus* sp. (Cyprinodontidae), *Nannostomus/Poecilobrycon* "pencil fishes," small loricariid catfishes, Gasteropelecidae, *Leporinus* sp. (Anostomidae), small pimelodid catfishes, and the electric eel (*Electrophorus*) (see Table 15).

KINGFISHER DENSITIES

TERRITORIAL BEHAVIOR

All five species of kingfishers seemed to be strongly intraspecifically territorial. Although aggressive acts were observed only rarely (100 times in 338 hours of direct observation), two conspecifics of the same sex were never observed within sight of each other without aggression between them. Although birds were not marked or banded, individuals recognizable by plumage or behavioral quirks were found consistently at the same spots month after month. When aggressive encounters were observed, they were often quite intense, with aerial pursuit chases averaging from 113 seconds in *amazona* to 3.3 seconds in *inda*. These bouts were almost always accompanied by harsh vocalizations. The extreme case occurred between two female *amazona* on Jaçana Lake on ISS on 1 October 1974: one chased the other around the lake continuously for 18 minutes; the contestants perched in nearby trees for 1 second after the aggressor had knocked its opponent out of the air and into the water for an instant; the chase then resumed, with intermittent periods of perching lasting from 10 to 218 seconds, for another 12 minutes before one finally drove the other away from the lake. Thus, kingfishers occasionally invest considerable energy in combat to determine territorial limits. Rigid territoriality in both breeding and nonbreeding season has been found in *Ceryle alcyon* (Davis 1982) and many Old World kingfisher species (Fry 1980a, Forshaw 1983).

Limited sample sizes prevented statistical examination of differences between inter- and intra-sexual aggression within each species. Qualitatively, individuals of the same sex never tolerated each other's presence, whereas individuals of different sexes (pairs?) showed more tolerance. However, males sometimes chased females and females sometimes chased males. Excluding the previously noted half-hour episode, there was no obvious trend in intensity of inter- vs. intra-sexual aggression.

Aggressive interactions between species were rarely recorded, i.e., only 40 instances in thousands of hours of field time and 338 hours of direct, timed observations (Table 5). Interspecific aggression usually involved only displacement of the victim from a perch and never the post-displacement aerial chases seen in intraspecific aggression; the 40 cases averaged 1.03 seconds (range 1 to 2) in duration. In contrast, intraspecific chases averaged 53.4 seconds (or 5.8 seconds, with a range from 1 to 20, if the extreme case in *amazona* noted previously is removed). Thus, interspecific aggression occurred much less frequently and did not last as long, once initiated, as intraspecific aggression. The harsh vocalizations that often accompanied intraspecific aggression were seldom noted in interspecific aggression.

TABLE 5. Number of aggressive interactions observed between kingfisher species.

Aggressor	Victim				
	torquata	*amazona*	*inda*	*americana*	*aenea*
torquata	32	12	0	0	0
amazona	0	37	3	9	0
inda	0	0	7	6	2
americana	0	0	0	16	8
aenea	0	0	0	0	4

Of the 40 cases of interspecific aggression, 29 were directed at the next-smaller species on the size scale, 11 at the species two size-classes smaller, none at species three or four size-classes smaller, and none at species larger than the aggressor (Table 5); the aggressor succeeded in displacing the victim in all 40 cases.

On numerous occasions, individuals of different species were noted hunting from the same tree without any sign of antagonism. I estimated that any kingfisher had at least one individual of another species in view at least 20% of its hunting time. Thus, the infrequency of interspecific aggression is even more striking, especially in light of the heavy overlap in diet between many species pairs (see Prey Size section). It is unknown what provoked the rare instances of interspecific aggression.

For one species, *inda*, more interspecific than intraspecific aggressive acts were recorded. All eight of these cases, however, occurred at a single spot on Lake Tumi Chucua favored by all kingfishers where a small stream emptied into the lake. Fish seemed particularly abundant at this site. All eight cases were displacements of smaller species from the single bare branch commanding the best view of this spot. Perhaps the other acts of interspecific aggression were the result of a similar situation: a favored hunting perch was occupied by another species.

In view of the frequency with which interspecific aggression and dominance hierarchies have been found among other species overlapping as heavily in resource use as these five kingfisher species (e.g., Willis 1966, Williams and Batzli 1979, Robinson 1981, Waite 1984, Caruthers 1986, Alatalo and Moreno 1987), the scarcity of observations of such behaviors among the kingfishers is noteworthy.

DENSITY SAMPLING METHODS

I attempted to measure densities of kingfishers and other fish-eating birds at the intensive sites and also at some other sites visited only occasionally. Because kingfishers are highly mobile within their large territories, results of a given census were greatly influenced by the amount of time spent per unit-length of shoreline. The censuses could not be standardized for differing amounts of time, because more time was spent in areas with more kingfishers. Therefore, the censuses should be regarded not as accurate, absolute-density data but as approximations, useful mainly for comparisons of relative abundance.

Another sampling problem was that kingfisher species were not all equally detectable. Few individuals of *torquata* or *amazona* were overlooked, but the three smallest species were easily missed. The most conspicuous of the three smallest was *americana*, because it preferred exposed perches. *Inda* and *aenea*, however, were difficult to see because, as has been noted by many observers (summarized by Forshaw [1983]), they prefer the shaded interiors of trees and bushes; *aenea* in particular was hard to detect. Furthermore, when flushed, *inda* and *aenea* tended to fly back away from the shoreline through the trees, retreating farther into the completely shaded, flooded shoreline, whereas the other three species flew out over the open water. I therefore have little confidence in rigorous comparisons of densities between species, particularly for *aenea*. I feel that density figures calculated for are disproportionately lower than the true densities of the other species. My general impression was that all five species were more or less equally common at Tucuchira and Tumi Chucua.

I measured shoreline distances by using the known length of a dugout canoe and counting the number of "dugout lengths" traversed while paddling along the shoreline. Because 100 m of stream has two perch-lined shores, i.e., both banks, whereas 100 m of lake shoreline has only one perch-lined shore, the census distances for stream sites were doubled for the density calculation unless the stream was so narrow that both shores could be fished from either bank (e.g., Tucuchira section E; see below). I assumed that kingfisher territory size is determined by the length of perch-lined shore; thus the linear distance of a kingfisher territory on a stream, with its two opposite but nearby shorelines, should be half of that on a lake, all else being equal. The energetics of defending a 200 m linear territory may be quite different from two parallel 100 m segments, but in the absence of relevant data, this problem has necessarily been ignored. The water current was too slow and the water depth too great for formation of the ripples found to influence fish densities and, therefore, densities of *Ceryle alcyon* by Brooks and Davis (1987).

Two factors may interact to influence kingfisher density: surface fish densities (methods and data in previous sections) and availability of hunting perches. I quantified

TABLE 6. Availability of kingfisher hunting perches. Study site abbreviations are: ISS = Isla de Santa Sofía; CL = Camungo Lake; JL = Jacana Lake; TC = Tumi Chucua; QT = Quebrada Tucuchira; AS = Arroyo Salsipuedes; and AR = Amazon River.

Perch Type	ISS (CL)	ISS (JL)	TC	QT (A)	QT (B)	QT (C)	QT (D)	QT (E)	AS	AR
foliated tree 1-10 ft. high	.087	.250	.707	.500	.654	.347	.885	.929	.931	.194
foliated tree 11-20 ft. high	.077	.300	.951	.685	.962	.389	.971	.929	.966	1.000
foliated tree 21-30 ft. high	.077	.350	.988	.667	.962	.389	.957	.929	.966	1.000
foliated tree 31+ ft. high	.077	.250	.829	.741	.923	.375	.943	1.000	.793	.896
leafless tree 1-10 ft. high	.053	.050	.012	.056	.038	.069	.057	.071	.069	.015
leafless tree 11-20 ft. high	.068	.075	.012	.018	.038	.083	.086	.000	.103	.044
leafless tree 21-30 ft. high	.063	.150	.012	.000	.038	.111	.043	.000	.103	.044
leafless tree 31+ ft. high	.072	.075	.012	.000	.038	.125	.043	.000	.034	.044
bush	.092	.175	.634	.296	.615	.306	.943	.857	.862	.194
snag	.034	.025	.024	.000	.077	.222	.043	.071	.484	.075
hidden tree or bush	.043	.175	.341	.000	.385	.222	.671	.286	.379	.015
recessed pool with tree 1-10 ft.	.019	.125	.037	.000	.038	.056	.014	.000	.034	.000
recessed pool with tree 11-20 ft.	.019	.125	.061	.000	.038	.056	.014	.000	.034	.000
recessed pool with tree 21-30 ft.	.019	.125	.061	.000	.038	.042	.014	.000	.034	.000
recessed pool with tree 31+ ft.	.014	.023	.061	.000	.038	.056	.014	.000	.034	.000
recessed pool with bush	.010	.100	.061	.000	.077	.042	.014	.000	.034	.000
recessed pool with snag	.010	.025	.012	.000	.000	.014	.000	.000	.034	.000
recessed pool with hidden perch	.000	.000	.048	.000	.077	.106	.014	.000	.000	.000
any perch 3-10 ft. in height	.164	.350	.939	.519	.750	.833	1.000	1.000	.965	.179
any perch higher than 10 ft.	.150	.600	.989	.925	1.000	.764	1.000	1.000	1.000	1.000
any perch higher than 20 ft.	.130	.525	.989	.925	.962	.736	.985	1.000	.965	1.000
no suitable perches available	.483	.375	.000	.074	.000	.166	.000	.000	.000	.000
number of sample intervals	207	40	82	54	26	72	70	14	29	67

perch availability by paddling along the shoreline of each intensive site and marking off each shoreline into 7.3 m intervals (2 dugout lengths). This distance was the maximum distance between perches that would allow the smallest kingfisher, *aenea,* to attack every meter of shoreline. It was empirically determined that from its mean perch height of 1.2 m, *aenea* would attempt to capture prey at lateral distances of up to 3.65 m from the perch and no farther. The larger the kingfisher, the longer this lateral distance, because of increasing mean perch height (see discussion in Dive Entry section).

Availability of various types of kingfisher hunting perches (see Table 11) was scored on a presence/absence basis for each 7.3 m interval. A perch did not have to be within the sample interval itself to be scored positively. For example, a single leafless tree 12 m high might result in a positive score for that perch type for 10 adjacent sample intervals, because a *torquata* using this perch could attack the shoreline at distances of up to 36.5 m in either direction. To score positively for all 10 sample intervals, the view from that perch had to be unobstructed by other vegetation (seldom the case), and the tree had to be on the effective shoreline. With increasing distance back from effective shoreline, the potential attack radius is reduced; thus, if that same leafless tree were 36.5 m back from effective shore, only 1 sample interval could potentially be attacked from that perch. The results, given in proportion of sample intervals in which each perch type was available, are presented in Table 6.

SEASONAL DENSITY DIFFERENCES

The ratios of low-water to high-water densities of kingfishers (Table 7) were moderately consistent among the study sites, especially in light of the small sample sizes and small areas involved at some sites. Low-water densities were substantially higher than high-water densities at all nine sites (Tables 8-10).

TABLE 7. Ratios of low-water to high-water season densities of kingfishers.

Site	Ratio (low-water: high-water)
Camungo Lake	4.0
Jacana Lake	4.8
Tucuchira, section A	—
Tucuchira, section B	4.6
Tucuchira, section C	2.1
Tucuchira, section D	6.0
Tucuchira, section E	2.2
Río Cayarú	3.4
Amazon River at ISS	4.1

Different habitats showed different patterns of seasonal change among size-classes of fishes (Fig. 4). Given this complexity, and given that each kingfisher species showed some differences in habitat preferences and size-classes of prey (both discussed in next section), it can be asked if the degree of seasonal change for each kingfisher population corresponded to the degree of seasonal change in its preferred fish size-classes in its preferred habitat. A good relationship (Fig. 6) exists between changes in fish and kingfisher populations (Spearman rank r_S = 1.00, P < .01). In other words, the species with the lowest degree of seasonal change in prey density, *aenea*, showed the lowest degree of seasonal change in diet (small fishes in shaded habitat). The species with the greatest seasonal change in density, *americana*, had the diet with the most drastic decline at high-water season (small dishes in lakes and streams). This strong correlation provides circumstantial evidence for the importance of surface fish density as a determinant of kingfisher density.

If a one-to-one correspondence existed between degree of seasonal change in fishes and their kingfisher predators — that is, if kingfisher populations tracked fish populations exactly — all points would have fallen along the diagonal line in Figure 6. But all the points lie above the line, indicating that kingfisher populations do not fluctuate as much as fish populations. This was expected, because kingfishers cannot match fishes in reproductive rate and generation time, nor are they highly migratory. Environmental carrying capacity for kingfishers may be set at an annual "bottleneck" period, probably the high-water season, when surface fish populations are at their lowest. The increase in kingfisher density at low-water did not keep pace with the increase in surface fish density, even though low-water kingfisher censuses included the annual recruitment of juveniles. These results imply that food was not a limiting factor at low-water season.

Do kingfishers affect fish populations? As discussed previously, no within-season changes in fish density were detected, but this absence of detectable change does not necessarily mean that kingfishers did not reduce fish populations significantly, because numbers lost to kingfisher predation could have been replaced by reproduction. Another possibility is that, rather than limiting surface fish population, kingfishers may limit the time spent by fishes on the surface. Except perhaps for the rare Gasteropelecidae, none of the kingfisher prey fishes was restricted to the upper 10 cm of water. An increase in kingfisher density might simply cause a shift in time spent on the surface by these fishes or an increase in alertness to surface predation. Surface fishes seemed to be extremely aware of movements above water, and an increase in diving frequency by kingfishers might disproportionately increase this awareness. Two shreds of evidence support this. First, kingfishers rarely return to the same spot after making a dive, successful or not: frequencies of returning to the same spot after a dive range from 4% to 13% for the five species. Second, experiments to measure time needed for surface fish density to recover after a splash (imitating a kingfisher strike) to pre-splash levels indicated that 5-10 minutes were necessary for fishes under 2 in., and at least 20 minutes for larger size-classes.

Perhaps kingfisher territory size is determined not so much by the number of fish within that territory as by an upper limit to the rate of diving above which surface fishes become too alert to predation from above. Kingfisher strike success must depend almost

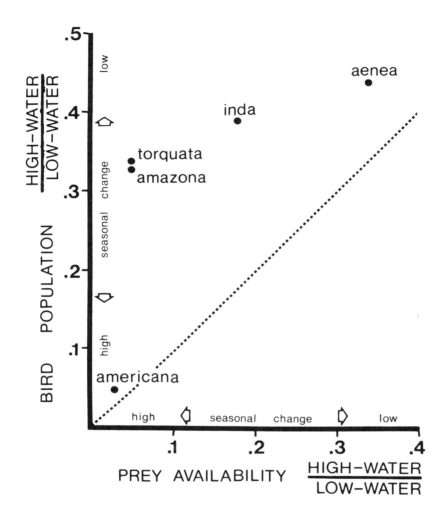

FIG. 6. Comparison of degree of seasonal change in kingfisher densities with degree of seasonal change in prey availability for each kingfisher. A high value indicates a low degree of seasonal change, and a low value, a high degree of seasonal change. The dotted line represents the line on which the points would fall if there were a 1 to 1 correspondence between change in prey availability and change in bird population.

entirely on surprise. No pursuit is involved; and once a dive is initiated, only minute alterations in course can be made, especially in the crucial last 2 m or so. If a fish sees a kingfisher coming, it can almost certainly escape. Thus, the optimum kingfisher strategy might be to lull its prey by maintaining a territory large enough so that the rate of striking per school is low, and by minimizing the number of dives necessary by taking as large a fish as possible. But a corollary of this is that the presence of other kingfisher species would constitute interference as well as exploitation competition. Each time that any kingfisher dives, it potentially increases the alertness of surface fishes to future surface predation. Therefore, it should benefit a territory-holder to drive out all smaller (i.e., beatable) kingfishers, regardless of species. Yet this did not occur.

If kingfisher carrying capacity is determined solely at high-water and food is not limiting the remainder of the year, all niche metrics taken at low-water are essentially irrelevant. However, there were no statistically significant differences in any niche parameters between the two seasons. There did not seem to be any feature of kingfisher foraging behavior that changed at high-water season. This contrasts with comparisons of feeding parameters between the five-species sites and the three-species site, Salsipuedes, where the reduced size of prey taken by *amazona* and *americana* was immediately obvious in the field. No such conspicuous change was evident between low-water and high-water within sites. The only parameters that obviously changed with a change in water level within sites were fish and kingfisher densities at high-water.

Where do the kingfishers go during the high-water season? The answer is straightforward for *aenea* and *inda*: the forest, when flooded, provides completely shaded habitat for these species. Forested land adjacent to rivers and streams, unsuitable kingfisher habitat at low-water season, becomes suitable. The populations of these two species presumably spread out to occupy this newly available habitat.

The answer is much less clear for *torquata*, *amazona*, and *americana*. The two largest species are found to some extent in open areas of flooded forest and in newly flooded open areas adjacent to rivers and streams. This seasonal increment in suitable habitat, however, did not seem to account for the degree of density reduction. A portion of their populations seemed to vanish completely from Quebrada Tucuchira and ISS. The case of *americana* is the most puzzling of all. This species disappeared completely from the Tucuchira-ISS area. Its absence at high-water season was conspicuous: the five-species system became a four-species system at this season. That the density value for *americana* at high-water season was not zero was due solely to three records of *americana* seen at Tucuchira in the last two weeks of high-water season. To what extent this disappearance is a regional phenomenon is not known. No *americana* were seen anywhere else in the region in high-water season in limited travel to other "quebradas" such as Arara and Guaçarí.

DENSITIES IN DIFFERENT HABITATS

Lakes (Table 8)

Camungo Lake. This lake, on ISS, had about 1500 m of shoreline and was characterized by extensive floating vegetation mats dominated by *Paspalum.* As a result, "effective" shoreline was from 3 to 100 m from true shoreline, so that trees and shrubs close enough to open water to be suitable hunting perches for kingfishers are in short supply.

Camungo Lake had the lowest availability of perches in almost every category (Table 6), dramatically so in the major categories "any perch 3-10 ft.," "any perch higher than 10 ft.," and "any perch greater than 20 ft." It also had the highest percentage (nearly 50%) of shoreline lacking any suitable perches. Perhaps even more significant was that those areas without suitable perches are clumped, creating large stretches of shoreline unsuitable for kingfishers. One such gap was 400 m long, more than a quarter of the lake's perimeter.

Camungo had the lowest density of kingfishers of any site except Tucuchira section A and the main channel of the Amazon River. Because surface fish densities (Table 2) were higher here than anywhere else except Jaçana Lake, the lack of suitable hunting perches almost certainly accounted for the low kingfisher densities. Because nearly 50% of the shoreline was unsuitable kingfisher habitat, calculating a density figure only for the length of shoreline with suitable habitat doubles the density, making it more comparable to that at other sites. Only three species were present in numbers; *inda* and *aenea* were rare or absent because of the rarity of shaded perches.

Jaçana Lake. "Jaçana Lake" is a small lake adjoining the downstream end of Camungo Lake. These two "lakes" were in reality connected by a narrow channel, even at the peak of low-water season. This lake's 140 m of shoreline was fringed by a nearly continuous floating vegetation mat dominated by *Paspalum.* The width of the mat varied from 1 to 120 m, wide enough to restrict availability of suitable hunting perches in all height categories. Only Camungo Lake, with its wider floating mat, had a lower percentage of shoreline with suitable perches. The Amazon River at low-water had a lower availability of perches 3-10 ft. high because the steep, muddy banks lacked brushy growth; at high-water the banks were inundated and water flowed under the undergrowth of the shoreline. Jaçana Lake was particularly rich in leafless trees, which are favored kingfisher perches.

Jaçana Lake, in contrast to adjacent Camungo Lake, had the highest kingfisher densities of any site except Tucuchira section B. Jaçana Lake also had by far the highest surface fish density of any site. When Jaçana Lake fish counts were separated from Camungo Lake's, surface fish densities were found to be 1.3 times higher in Jaçana Lake and 2.1 times higher for fish larger 2 in., the size class most heavily used by the two largest species, which formed the majority of the Jaçana Lake kingfisher population (72% at low-water, 100% at high-water).

Lake Tumi Chucua. Lake Tumi Chucua is a large, several km long, oxbow lake adjacent to the Río Beni, Bolivia. One stretch along the southeastern shore about 600 m long was censused 11 times, and a similar section along the western shore about 700 m long was censused only 3 times. Neither was surveyed during high-water season. Lake Tumi

TABLE 8. Densities of piscivorous birds (individuals/km shoreline) at lake study sites by season. An "X" indicates that that species was recorded at that site but not on a census. See Table 6 for abbreviations for study sites. Abbreviations for seasons are "LW" = low-water and "HW" = high-water.

	Study Site and Season						
	ISS (CL)		ISS (JL)		TC (SE)	TC (W)	
Species	LW	HW	LW	HW	LW	LW	
Ceryle torquata	0.95	0.24	5.59	1.65	2.18	2.29	
Chloroceryle amazona	0.95	0.24	5.59	1.65	2.36	2.57	
Chloroceryle inda	—	—	X	—	2.00	1.71	
Chloroceryle americana	0.38	—	3.73	—	2.36	2.29	
Chloroceryle aenea	—	—	0.93	—	1.09	1.43	
all kingfishers	2.28	0.48	15.84	3.30	9.99	10.29	
Busarellus nigricollis	0.43	X	4.97	0.55	X	0.57	
Pandion haliaetus	0.33	X	0.62	—	0.72	0.86	
Phaetusa simplex	0.95	X	2.17	2.75	—	—	
Sterna superciliaris	0.48	—	0.93	—	0.36	0.29	
Butorides striatus	0.24	1.59	0.93	6.04	2.36	2.29	
Casmerodius albus	0.43	1.10	1.86	3.30	0.18	0.29	
Ardea cocoi	0.24	—	X	X	—	X	
Pilherodius pileatus	0.29	—	—	X	—	X	
Agamia agami	0.05	—	0.31	—	X	—	
Tigrisoma lineatum	0.05	—	—	—	—	—	
Phalacrocorax brasilianus	X	—	—	—	0.36	0.57	
Anhinga anhinga	0.05	—	—	—	X	0.29	
all piscivorous birds	5.82	3.17	27.63	15.94	13.97	15.45	
number of censuses	14	5	23	13	11	7	

Chucua differed from Camungo and Jaçana lakes in lacking continuous floating vegetation mats. Small patches of floating vegetation were scattered through both sections, but nowhere did they restrict availability of suitable kingfisher hunting perches. Perches of all heights were nearly continuously available along the shore (Table 6, quantified for southeastern section only). In this respect, the shoreline of Lake Tumi Chucua resembled the stream sites much more closely than the lakes at ISS: shaded perches were found along 33.5% of the shoreline, contrasting with the much lower values for Jaçana (17.5%) and Camungo (4.3%) lakes.

Total kingfisher density at Lake Tumi Chucua was higher than at any other site except Jaçana Lake and Tucuchira sections B and D. Densities of *aenea* on both sections were higher than at any other site.

Streams (Table 9)

Quebrada Tucuchira, section A. This section was the last 200 m of the stream before it emptied into the Amazon River. An extensive mat of floating vegetation covered most of this section, except for a central channel 0.5 m wide. The mat was so wide on one side of the stream that there were no trees close enough to effective shoreline to be suitable kingfisher hunting perches. Perches higher than 10 ft. were present along 92.1% of the shoreline (Table 6), but perches under 20 ft. were much more restricted (52.6%). In its reduced availability of perches and extensive floating vegetation mat, this section was atypical for stream shoreline, resembling more closely the lakes on ISS.

This section had the lowest kingfisher densities of any site except the Amazon River, which section A resembled in lacking small kingfishers. It was not used at all by kingfishers at high-water season because density of surface fishes was low (Fig. 7) and because perch availability was reduced, especially for perches 3-10 ft. high. Also, the high content of suspended silt in the water, similar to the Amazon River in being nearly opaque, may have reduced surface fish density. This turbidity was produced by a "white-water" tributary stream that emptied into Tucuchira at the junction of sections A and B; also, during periods of low run-off, "white-water" from the Amazon actually penetrated upstream to this point.

Quebrada Tucuchira, section B. This was a 100 m section lines with tall (up to 30 m), flooded forest on both shores. The channel varied in width from 8 to 15 m. A few patches of *Paspalum* were scattered along the shore, reducing the availability of low perches in some spots. Counterbalancing this was a relatively high availability of snag perches (Table 6), which were especially favored by *americana*. Otherwise, this section was typical of small streams throughout the region.

Kingfisher densities here were higher at either season than at any other site except Jaçana Lake: two species, *americana* and *inda*, were particularly common, probably because surface fish densities were high and availability of their favored hunting perches was nearly unlimited.

Quebrada Tucuchira, section C. This section of Tucuchira, estimated to have about 200 m of shoreline, had extensive floating vegetation mats. The stream widened to 50 m in this

TABLE 9. Densities of piscivorous birds (individuals/km shoreline) at stream study sites by season. See Table 8 for explanation of codes.

	\multicolumn{11}{c}{Study Site and Season}										
	QT (A)		QT (B)		QT (C)		QT (D)		QT (E)		AS
Species	LW	HW	LW	HW	LW	HW	LW	HW	LW	HW	LW
Ceryle torquata	0.83	—	5.21	1.92	1.94	1.42	3.00	0.67	—	—	0.67
Chloroceryle amazona	0.83	—	3.96	0.77	1.67	1.79	2.50	0.67	—	—	0.78
Chloroceryle inda	—	—	3.75	0.77	1.39	X	2.50	1.34	3.64	2.50	—
Chloroceryle americana	X	—	5.63	—	1.39	—	3.00	0.67	—	—	1.50
Chloroceryle aenea	—	—	1.04	0.77	0.28	X	1.00	0.67	0.91	X	—
all kingfishers	1.66	—	19.59	4.23	6.67	3.21	12.00	4.02	4.55	2.50	2.95
Busarellus nigricollis	—	—	X	—	X	—	0.50	X	—	—	X
Pandion haliaetus	0.21	—	—	—	—	—	0.25	—	—	—	—
Phaetusa simplex	—	—	—	0.77	2.22	1.07	X	—	—	—	—
Sterna superciliaris	—	—	—	—	—	0.71	—	—	—	—	—
Butorides striatus	0.21	3.93	0.63	2.31	1.11	1.43	0.25	1.00	—	—	0.89
Casmerodius albus	—	—	—	—	—	—	X	—	—	—	0.06
Ardea cocoi	—	—	—	—	—	—	X	—	—	—	0.22
Pilherodius pileatus	—	—	—	—	—	—	—	—	—	—	X
Agamia agami	—	—	X	—	—	—	X	—	0.91	—	—
Tigrisoma lineatum	—	—	—	—	—	—	X	—	—	—	—
Phalacrocorax brasilianus	—	—	—	—	—	—	—	—	—	—	—
Anhinga anhinga	—	—	—	—	—	—	X	—	—	—	0.22
all piscivorous birds	2.08	3.93	20.22	7.31	10.00	6.42	13.00	5.02	5.46	2.50	4.34
number of censuses	24	14	24	13	18	14	8	3	11	8	18

section, forming an expanse of open water almost pond-like in aspect. Because of the wide fringe of floating vegetation, perch availability was low compared to the rest of Tucuchira and was more similar to Jaçana Lake than any other site.

Kingfisher densities in this section were lower than in more typical stream sections such as B and D, primarily as a consequence of reduced perch availability.

Quebrada Tucuchira, section D. Section D was a 250 m section lined continuously on both banks with tall, flooded forest. *Paspalum* mats were absent. Shaded perches were relatively more common here than at any other site (except the canopy-closed stream, section E, where all perches were of this type by definition). This section of Tucuchira was the most representative of typical stream shorelines of the region.

Kingfisher densities here were high and approximately equal among species. Surface fish densities here were not as high as in areas such as Jaçana Lake and section B, which also had higher kingfisher densities (Fig. 7). The absence of *Paspalum* mats was presumably responsible for the lower fish densities.

Quebrada Tucuchira, section E. This 100 m segment was actually a smaller tributary of Tucuchira. The channel was only 2-3 m wide and was completely shaded by forest canopy that closed overhead; thus, all perches were classified as "shaded." *Paspalum* mats were absent. This was a typical, small, forested stream of the region.

Such closed-canopy streams represent a distinct habitat-type for kingfishers: three of the five species were virtually never found in this habitat, whereas *inda* reached maximum densities here. *Inda* and *aenea* are characteristic of these shaded, forest streams. Total kingfisher density was low relative to other stream habitats, probably because surface fish availability was reduced (Fig. 7) and because appropriate habitat for three kingfisher species was absent.

Arroyo Salsipuedes. A 100 m section of Arroyo Salsipuedes (Río Yata, Bolivia) was censused regularly. The stream channel was 3-5 m wide and was clogged with brush and snags; *Paspalum* mats were absent. Perches of all types and heights, including shaded perches, were available nearly continuously (Table 6).

Kingfisher densities were lower here than at any other stream site except section A of Tucuchira. Reduced density of surface fishes (Fig. 7) was almost certainly the cause.

Comparison of densities of kingfishers at my stream sites with those found for *Ceryle alcyon* on two streams in North America (Davis 1982, Brooks and Davis 1987) shows that densities at my sites were much higher, not just for total density of all kingfishers but also on a per species basis. My two sites that were most comparable in stream width to the streams in Pennsylvania and Ohio studies by Davis (1982) and Brooks and Davis (1987) were Tucuchira B and D. To compare the two data sets, I divided my density figures by one-half (because my density calculations counted opposite banks of the same stream separately) and doubled the density data of Brooks and Davis (because they used number of pairs, not number of individuals). The temperate zone sites had densities (0.3 and 1.0 kingfishers per km, respectively) substantially lower than those at my tropical sites (9.8 and 6.0 kingfishers per km during low-water, and 2.1 and 1.0 per km during high-water; Table 9).

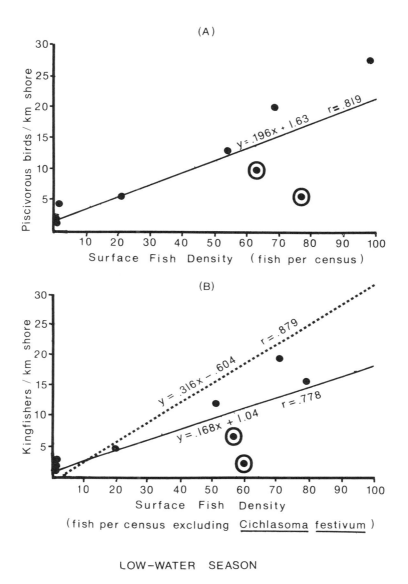

FIG. 7. Relationship between surface fish density and bird density for each study site at low-water season. *Cichlasoma festivum* densities were excluded from the upper figure because this species of fish is seldom, if ever, taken by kingfishers (see text). The two circled points represent Camungo Lake and Tucuchira section C. The dotted lines represent the regression line if kingfisher density is corrected for perch availability.

Rivers

Results of the river censuses are presented in Table 10. Because the maps available for the region were crude, the distances covered by the censuses could not be measured accurately, particularly because the river courses were typically more tortuous than shown on maps.

Río Cayarú. The Río Cayarú, a small river in Dpto. Ucayali, Peru, emptied into the Amazon River near ISS. Its channel was 10-30 m wide. The banks were densely forested, thereby providing kingfisher hunting perches of all types, including shaded perches, nearly continuously.

Kingfishers were censused five times from the mouth of the river to the village of Bella Vista, estimated to be 15 km from the mouth (= 30 km of riverbank). Densities were much lower than in stream habitats, particularly for the three smallest species.

Río Javarí. The Río Javarí, a moderately large tributary of the Amazon River, forms the boundary between Dpto. Ucayali, Peru, and Amazonas, Brazil. Channel width was estimated to be 300-400 m in the census area, from the Brazilian town Attalaia do Norte to the Peruvian village Pobre Alegre. The banks were forested almost continuously, and availability of hunting perches was unrestricted, except for reduction in perches under 10 ft. during low-water season in areas with steep banks.

Kingfisher densities here were lower than at any other site. The three smallest species were completely absent, which is typical for the main channel of the Amazon River. In over 100 hours of travel time on the Amazon, I never observed a feeding individual of any of the three smallest species. Also, *torquata* always outnumbered *amazona* on the Amazon; this contrasted with all other sites, where densities of these two species were consistently similar.

Amazon River, ISS to Puerto Nariño (section B). On Sept. 20, 1974, I censused the Colombian bank of the Amazon from ISS to the town of Puerto Nariño, a distance estimated to be 50 km. Kingfisher densities were low, and only two species, *torquata* and *amazona,* were present; this was typical for the Amazon River.

Differences Between Rivers and Streams

In river and stream habitats, a gradient of decreasing density and diversity extends from streams to large rivers. The larger the river, the less likelihood of its having the three smallest species, *aenea, inda,* and *americana,* and the lower will be the density of the two largest species, *torquata* and *amazona*. Also, as the river increases in size, the ratio of *torquata* to *amazona* becomes greater. Thus the gradient proceeds from a stream such as Tucuchira, with all five species in more or less equal abundance; to a small river such as the Cayarú, with all five species but in reduced densities; to a medium-size river such as the Javarí, with only three species, the smallest very rare, a further reduction in density, and an increase in the ratio of *torquata* to *amazona;* to a large river such as the Amazon, with only the two largest species remaining. Although I had census data only from one example of each type of river, qualitative observations on several other streams and rivers in Amazonian Colombia, Peru, and Bolivia supported generalizations from the census data.

TABLE 10. Densities of piscivorous birds (individuals/km shoreline) at river study sites. Abbreviations for seasons are "LW" = low-water, "HW" = high-water, and "HW-LW" = transition period from HW to LW. An "X" indicates that the species was recorded there, but not on a census. A "+" indicates that the species was recorded on censuses but at mean densities lower than 0.05 individuals/km shoreline.

	Study Site and Season						
	Río Cayarú		Río Javarí		Amazon (A)	Amazon (B)	
Species	LW	HW-LW	LW	HW-LW	LW	LW	HW
Ceryle torquata	2.7	0.6	0.2	0.3	0.2	0.2	X
Chloroceryle amazona	1.0	0.5	0.1	0.1	0.1	0.1	—
Chloroceryle inda	0.2	—	—	—	—	—	—
Chloroceryle americana	0.2	0.1	—	+	—	—	—
Chloroceryle aenea	—	—	—	—	—	—	—
all kingfishers	4.1	1.2	0.3	0.4	0.3	0.3	X
Busarellus nigricollis	0.1	0.1	—	—	—	—	—
Pandion haliaetus	—	+	—	—	X	—	—
Phaeusa simplex	0.3	0.8	0.1	0.3	0.2	0.4	0.3
Sterna superciliaris	—	—	+	+	+	—	—
Butorides striatus	0.3	0.9	+	+	+	X	0.1
Casmerodius albus	0.1	+	+	—	—	—	—
Ardea cocoi	0.1	0.1	+	+	—	—	—
Pilherodius pileatus	+	—	+	+	—	—	—
Agamia agami	+	—	—	—	—	—	—
Tigrisoma lineatum	—	—	—	—	—	—	—
Egretta thula	—	—	+	—	X	—	X
Phalacrocorax brasilianus	—	0.1	+	—	—	—	—
Anhinga anhinga	—	0.2	—	—	—	—	—
all piscivorous birds	5.0	3.4	0.8	1.0	0.5	0.7	0.4
number of censuses	2	3	2	2	5	7	15

Diamond and Terborgh (1967) observed a segment of this pattern of kingfisher distribution on their survey of a river in Amazonian Peru.

Although perch availability declines along this gradient, it cannot be completely responsible for the gradient in kingfisher density and diversity. Only on the largest rivers does perch availability decline to the point of influencing kingfisher density (through lack of shaded perches for *inda* and *aenea*). Even on the Amazon itself, the availability of low perches is about the same as on Jaçana and Camungo lakes, where *americana* is found in moderate densities; yet the Amazon lacks *americana* entirely. Availability of low perches on the Javarí, although not quantified, is undoubtedly much greater than on the two lakes at ISS; yet the density of *americana* was an order of magnitude lower.

Surface fish density on rivers was quantified only on the Amazon at ISS; here, fish density was the lowest on any site measured. This may have been caused by the increased amplitude of water-level fluctuations with increasing channel width, and the increasing silt content and current strength. The waters of the Amazon and Javari are virtually opaque, and this must hinder the ability of kingfishers to detect prey. Also, many surface fishes seem to be visual foragers, and the strong reduction in visibility may reduce their densities. Furthermore, the size distribution of surface fishes shifts toward larger size-classes, leaving fishes under 2 in. extremely scarce. This would certainly have a negative effect on at least the three smallest kingfishers.

INTERACTION OF PERCH AVAILABILITY AND FISH AVAILABILITY

Throughout the discussion of kingfisher density in various habitats, the effects of surface fish density and hunting-perch availability have been postulated as the main determinants of bird density. Surface fish density and perch availability, however, interacted in a conflicting manner. The microhabitat richest in surface fishes, the floating vegetation mat, was the primary agent of reduction of availability of hunting perches by extending the distance from shoreline trees to the effective shoreline. Thus, the habitats with the highest fish densities, associated with extensive development of the floating vegetation mat, tended to be precisely those habitats with the greatest reduction in perch availability (e.g., Camungo Lake, Jaçana Lake, Tucuchira section C). Optimum kingfisher habitat would thus contain patches of floating vegetation sufficiently small and scattered that perch availability was the site with the highest total kingfisher densities (except for the enigmatic Jaçana Lake).

A high positive correlation exists between bird density and surface fish density for the nine intensive sites for which both were quantified (Fig. 7). The effect of fish density on bird density seems to be linear over the ranges of values involved. Two points, circled on the graph, fall over far below the regression line; their bird densities were below levels predicted by their fish densities. These two points represent the tow sites, Camungo Lake and Tucuchira section C, with the greatest reduction in perch availability for their respective habitat types, lakes and streams.

How do we control for the effects of perch availability? If bird densities are calculated for the shoreline distance that contained any suitable kingfisher perch (i.e., removing the distance with no suitable perch), effect of reduced perch availability are partially removed.

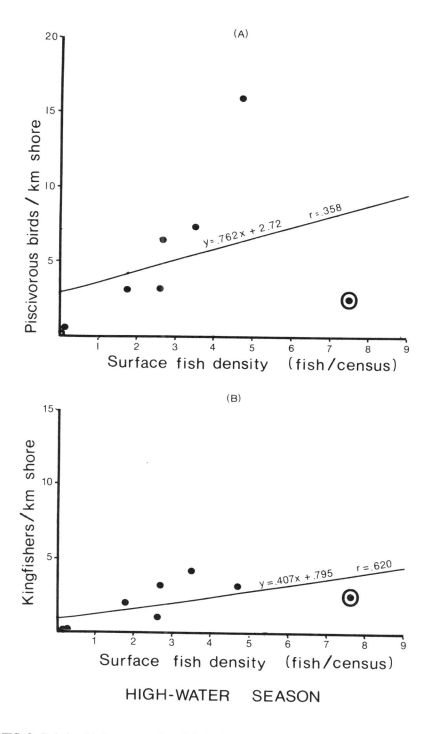

FIG. 8. Relationship between surface fish density and bird density at high-water season. The circled point represents Tucuchira section E. The dotted lines represent the regression if that study site is excluded from the calculations.

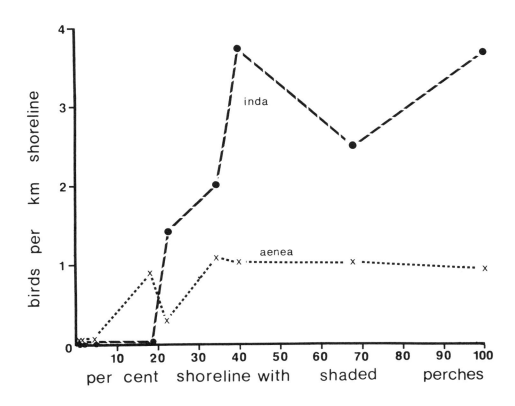

FIG. 9. Relationship between availability of completely shaded perches and density of *Chloroceryle inda* and *C. aenea*.

This can be further refined for the remaining shoreline by calculating densities only for the shoreline in which preferred habitat or hunting perches for particular species were available, i.e., for *inda* and *aenea* only for that distance of shoreline on which shaded perches or recessed pools, was available; for *americana*, for that distance on which any perch 3-10 ft. in height was available; and for *torquata*, for that distance on which any perch higher than 10 ft. was available. For *amazona*, the species with the correction factor is necessary. Using these new densities, adjusted for availability of suitable perches, the correlation coefficient of the new regression line (dotted line in Figure 7b) increases from .778 to .901.

At high-water season, surface fish density and kingfisher density both decrease. At this season, there is again a strong correlation between bird density and fish density (Figure 8b). At high-water, sites with low perch-availability do not fall below the regression line as they did at low-water. Apparently fish density is so low at high-water that it, rather than perch availability, becomes the factor that limits kingfisher densities. There was also another minor factor contributing to the decline in importance of perch availability: higher water levels brought effective shoreline closer to shoreline vegetation, increasing availability of hunting perches.

The one point in Figure 8b that falls farthest from the line, circled on the graph, is Tucuchira section E, the small, closed-canopy stream. Here kingfisher density fell far below that predicted by its fish density. I have no explanation for why this should be so. Also, when all piscivorous birds are included in the high-water calculations (Fig. 8a), the regression line is not significantly different form a horizontal line unless Tucuchira section E is removed form the calculations.

Examination of the relationship between availability of shaded perches (hidden perches plus any perch on canopy-closed streams) and densities of *inda* and *aenea* (Figure 9) shows that neither species is present at a site until at least 20% of the shoreline has shaded perches.

In summary, the interplay of prey availability and perch availability seems to determine kingfisher densities within the variation imposed by differences in season and habitat. Thus, prey availability and perch availability interact in a manner similar to that found by Janes (1984) for the Red-tailed Hawk (*Buteo jamaicensis*).

KINGFISHER FEEDING BEHAVIOR

STUDY METHODS

Whenever a kingfisher as encountered, it was followed for as long as possible with a dugout canoe (or by foot along the shore at Arroyo Salsipuedes), and the following data were recorded: (1) *species*; (2) *sex:* all five species are sexually dimorphic in plumage patterns, and the differences are obvious in the field (there were some indications of sexual differences in feeding behavior, but these were beyond the scope of the present paper); (3) *date*; (4) *time of day*; (5) *habitat*; (6) *perch type*; (7) *perch height*; and (8) *horizontal perch location*, with respect to distance from "effective" shoreline.

If a dive was initiated from the hunting perch, the following additional data were recorded: (9) *dive angle*, the angle made by the path from the hunting perch to the entry point into the water; (10) *distance of dive entry point from "effective" shoreline*; (11) *dive outcome*: miss or catch; (12) *prey type*; (13) *prey size*; and (14) *sunniness* and (15) *smoothness* of water at dive entry point.

When a bird moved from a hunting perch, the following information was also recorded: (16) *location of subsequent perch* (1976-77 only) with respect to whether the next hunting perch overlooked the same patch of water as the previous perch; and (17) *reason for moving*, i.e., why the bird left the hunting perch — for example, did it chase or was it chased by a conspecific or another species, or did it make a dive, or did it switch perches without diving, seemingly giving up? Additionally, (18) *time-activity data* were recorded for each perch, using a stopwatch to measure time spent in various activity categories.

A tape-recorded and field notebook were used to record information. Data for a given hunting perch were often incomplete. For instance, it was not always possible to determine prey type or prey size, and occasionally the outcome itself could not be seen for a given dive; also, various other parameters, such as dive angle or perch measurements, could not always be determined. Data in categories 9, 11, and 14-18 will not be discussed in detail here but will be presented elsewhere (Remsen, in prep.). Details of the methodology for the remaining categories will be discussed below.

Because several data categories relied on visual estimates of distance, and because I did not feel comfortable estimating these in metric units, the unit of "feet" was used and is retained in the presentation of data herein. Conversion to the metric system would have resulted in awkward, nonsensical data divisions, e.g., a "6.4-7.6 m" perch-height category for 21-25 ft.

Because the fish that kingfishers feed to their young are much smaller than those eaten by the adults, care was taken to eliminate from the data those fish taken to the nest. Prey and the preceding foraging bouts were included in the analysis of feeding behavior only if the prey was swallowed by the adult. This is not meant to downplay the importance of a critical phase of kingfisher biology, but to restrict the breadth of the questions investigated. For brief comments on the timing of breeding in the ISS-Tucuchira area, see Remsen (1978). (Skutch [1957, 1972] has studied the nesting biology of *torquata* and *amazona*, but *inda*, *americana*, and *aenea* have not been studied in detail; see Forshaw [1983] for summaries of anecdotal breeding information for these species).

QUALITATIVE DESCRIPTION OF FEEDING BEHAVIOR

All five kingfisher species hunt and feed in basically the same way. A hunting bird perches on a branch with gaze fixed on the water for long periods of time. The larger the species, the longer the mean time spent per perch (Willard 1985; Remsen, unpublished data). The intensity of the vigil of the hunting bird varied. At one extreme were cases in which the hunter leaned forward with tail raised quickly and briefly to the horizontal plane and wings held slightly away from the body. Although this posture was not always followed by a dive, it seemed obvious that the bird was preparing to dive. At the other extreme, the kingfisher's posture was more hunched, without any forward lean or tail-flicking and with wings against the body. The only distinction between this "relaxed" hunting pose and the "resting" category of time-activity data was that the bird was definitely watching the water (and occasionally would suddenly initiate a dive from this position).

Once a dive was initiated, the kingfisher flapped once or twice quickly as it left the perch, and with wings held slightly outward from the body, plummeted through the air toward the target. When the target was sufficiently far away laterally, the bird would fly with shallow wingbeats during the final few meters. Most dive paths were more-or-less the diagonal between the hunting perch and dive entry point, although more convex paths were not infrequent, allowing the kingfisher's final approach to be closer to the vertical than a straight diagonal approach.

Kingfishers almost never disappeared completely below the water, or if at all, only for a few centimeters. Because their momentum would seem to carry them well below the surface despite their high buoyancy, some braking action upon impact must be used, most likely with the wings. Thus, fish were captured only within the first few centimeters of the surface. Because refraction of light by water increases the difficulty of judging the true location of an object with increasing depth below surface, capture success probably diminished with decreasing angle of the dive, because refraction effects would increase sharply with decreasing angle. Yet success actually increased with decreasing dive angle (Remsen, in prep.).

After entry into the water, the bird left the water seemingly as quickly as possible, usually within 1 second of impact. No underwater pursuit of prey was ever observed. At my study site, this rapid escape from the water would certainly have been advantageous. Large aquatic predators, mainly fishes, were common and were attracted to splashing sounds. In fact, local fishermen splashed the water near their fishing lines to attract large fishes (as do fishermen elsewhere in Amazonia [Goulding 1980]). Although I did not witness predation on diving kingfishers by aquatic predators, a few local residents whom I considered reliable concerning natural history information claimed to have witnessed such an event. The rapidity with which any organic object was snatched from the water surface attested to the likelihood of such predation, as does the dependence of many large Amazonian fishes on fruit, leaves, and invertebrates that fall into the water (Goulding 1980). Fast-swimming, predatory, pimelodid catfishes, piranhas, other large predatory characins, and perhaps small caimans would be likely predators. Goulding (1980) found single birds in the stomachs of *Osteoglossum bicirrhosum* and *Serrasalmus rhombeus*, two fish species common at my study sites.

After a dive, the kingfisher almost always moved to a new perch overlooking a new patch of water. The disturbance created by the dive probably alerted small fishes in the area to surface predation, and it was then presumably more profitable to move on and let the patch "recuperate." After a successful dive, the bird flew to a perch and always manipulated the fish into a head-first position before swallowing. Larger prey items were beaten against a branch until they ceased to struggle. I never saw a kingfisher swallow even the smallest prey item in flight, although this has been reported for *Ceryle rudis* (Douthwaite 1971, Whitfield and Blaber 1978, Johnston 1989). Bouts of preening were most frequent immediately after a dive.

HABITAT PREFERENCES

Each foraging observation at the intensive sites was placed in one of the following habitat categories: (a) shorelines of lakes or streams adjacent to open water; (b) recessed pools, i.e., patches of open water isolated superficially from the main body of open water by floating vegetation or land, and adjacent to either lakes or streams; (c) "hidden" trees and bushes adjacent to open-water habitats, i.e., those areas in which the water flowed in underneath the shoreline vegetation, providing kingfisher perches shielded from the open sky and open water; (d) flooded forest, either permanently or seasonally; (e) canopy-closed streams, i.e., streams small enough that the canopies of opposite banks close over the stream, screening out most sunlight. Time spent searching for kingfishers was apportioned roughly equally among the various habitat categories, but differences in detectability biased the results toward underestimation of kingfisher densities in shaded habitats. Thus, absolute percentages must be interpreted with caution.

In the five-species community, habitat preferences (Table 11) of all five species are significantly different from each other (Chi square, $P < .001$) except for *torquata* vs. *amazona*. In the three-species community, all species are significantly different from each other (Chi square, $P < .05$), although it is doubtful that such small differences are biologically significant.

TABLE 11. Kingfisher habitat preferences. Data are presented as percentage of total number of observations of that species.

	Open Habitats		Shaded Habitats			
	open shorelines	recessed pools	hidden perches	flooded forest	closed-canopy streams	N
Five-species community						
torquata	87.2	9.9	2.2	0.7	—	695
amazona	84.9	12.3	2.6	0.3	—	962
inda	6.1	39.4	26.1	8.2	20.2	356
americana	58.5	31.2	9.0	0.3	1.0	992
aenea	2.5	52.5	33.1	7.6	4.1	194
Three-species community						
torquata	99.5	0.5	—	—	—	126
amazona	91.4	1.2	7.4	—	—	283
americana	94.1	3.3	2.7	—	—	908

Three species, *torquata*, *amazona*, and *americana*, were found primarily in open, sunny habitats (categories a and b above) and only occasionally (2.9–10.3% of all observations) in shaded habitats (c, d, and e). As noted by Slud (1964) and Stiles and Skutch (1989) in Costa Rica, the other two species, *inda* and *aenea*, predominated in shaded habitats, and when they did occur in open habitat, they usually frequented recessed pools, appearing only rarely (6.1% and 2.5% of all observations) at open edges of streams and lakes. *Inda* was found on closed-canopy streams to a greater extent than *aenea*; otherwise their use of shaded habitats was similar. Limited observations at small forest streams (< 2 m wide) indicated that *aenea* was more common there than *inda*.

Thus, habitat preferences group the five species into two categories: three species found mainly in open habitats and two mainly in shaded habitats. Substantial ecological separation between the two groups results is not as sharp as that found in many groups of birds, because all five kingfisher species occur together along the banks of streams such as Tucuchira and forested oxbow lakes such as my site at Lake Tumi Chucua and Willard's (1985) site in Amazonian Peru (and presumably elsewhere throughout Amazonia), where they hunt for fish mainly within 2 m outwards in either direction from the same shore, and because fish prey frequently moved back and forth between open and shaded water (personal observation).

In the absence of *inda* and *aenea* in the three-species community, the other three species did not expand into shaded habitats. If the reduction in surface fish densities in shaded habitats at Salsipuedes was proportional to the reduction in open habitats, fish densities may have been so low that hunting was unprofitable in shaded habitat there.

PERCH TYPE

Each hunting perch used by a kingfisher was placed in one of the following categories: (a) densely foliated tree (view of water obstructed by leaves except when bird perched on outer perimeter of the tree); (b) sparsely foliated tree (leaves not dense enough to obstruct view of water in most directions); (c) leafless tree; (d) leafy bush; (e) leafless bush; (f) snag (differs from leafless bush in having only one or two bare, prominent branches); (g) "hidden" tree (differs from other tree categories in that the kingfisher is not perched on the edge of open water with sky flows in underneath the tree, i.e., the kingfisher is shielded by a screen of foliage); (h) "hidden" bush (similar to "g" but a bush rather than a tree); (i) air (kingfishers occasionally hunted by hovering).

Most differences in perch-type preferences between species (Table 12) are determined by differences in perch-height preferences (see next section) and the unequal height distribution of perch types. For example, the larger the kingfisher, the fewer observations at low perches, i.e., bushes and snags, and the smaller the kingfisher, the few observations at high perches, i.e., trees.

Within broad height categories, kingfishers showed some perch selectivity. Leafless trees and snags, presumably because they afforded in unimpeded view, were used disproportionately. For example, snags were present in only 2.4-7.7% of the shoreline sample intervals at all intensive sites except Tucuchira section C and Salsipuedes (Table 6), yet frequency of use by the three smallest kingfishers ranged from 21.2% to 51.7% in the five-species community. One species in particular, *americana*, concentrated on snag perches. Such a preference for perches in bare trees and bushes is known for many bird species that hunt from perches, especially raptors (e.g., Bohall and Collopy 1984). Only two kingfisher species, *inda* and *aenea*, used hidden perches to any great extent. If weighted by the amount of time spent on the perch, use of these hidden perches increased by 10 percentage points. Willard (1985) also found that *inda* and *aenea* favored such types of perches.

In the three-species system, there was no shift toward lower perch types, e.g., bushes and snags, despite the significant downward shift in parch heights (see next section). This lack of a shift merely reflected the lower height of trees at Salsipuedes. Differences within species between the five- and three-species reflected minor changes in habitat structure rather than any difference in active selection of perch type by the kingfishers.

TABLE 12. Perch use by kingfishers. Data are presented as percentage of total number of observations of that species.

	Perch Type								
	densely foliated tree	sparsely foliated tree	leafless tree	leafy bush	leafless bush	snag	hidden tree	hidden bush	air	N
Five-species community										
torquata	29.9	33.7	28.6	1.1	1.4	2.6	2.5	—	0.3	653
amazona	28.0	31.0	15.1	2.4	0.4	18.8	2.7	0.2	1.3	914
inda	9.4	10.6	15.3	4.1	5.9	21.2	24.1	7.6	1.8	340
americana	3.9	6.8	12.2	6.2	8.7	51.7	6.8	2.9	0.6	964
aenea	4.3	3.2	3.2	18.3	2.2	29.0	30.1	9.7	—	186
Three-species community										
torquata	39.0	61.0	—	—	—	—	—	—	—	118
amazona	32.0	50.2	0.7	2.9	0.7	4.4	6.5	1.8	0.7	270
americana	7.1	9.3	0.4	9.0	14.8	56.2	1.7	0.8	0.6	964

PERCH HEIGHT

The height of each hunting perch was estimated to the nearest foot (0.31 m). Practice with objects of known height and placement of height markers at key positions on the intensive sites improved the accuracy of these estimates. If errors were made, they were probably in the direction of underestimation. Perches higher than 30 ft. were estimated only to the nearest 5 ft.

All distributions of hunting-perch heights for the five- and three-species communities (Figs. 10 and 11) are significantly different from one another, between species in the same community and within species in different communities (Kolmogorov-Smirnov test, P < .001) for all pairs, except *aenea* vs. *americana* in the five-species community, P < .01). Thus, all species were selecting different perch heights within a community; and the three species at Salsipuedes all shifted downward to lower perches from their preferred perch heights in the five-species communities. As found by Willard (1985), the larger kingfisher, the higher the perch. However, Willard's (1985) data on perch heights for kingfishers in Amazonian Peru differ greatly from mine in that all species but *aenea* used perches much higher above water. The greatest difference between our data is for *inda*, for which Willard's mean perch height was approximately 18 ft., versus 5.5 ft. from both studies. I believe that these differences are real and not the result of differences in our estimation procedures. Whether these differences are caused by different availabilities of perches, by differing fish size distributions and, therefore, different optimal perch heights, or by other factors is not known.

Why do different kingfisher species select different perch heights? In general, perches were not a limiting resource, and so differences in perch heights did not result in any direct ecological separation. An upper limit to perch height is set presumably by the constraints of target visibility — the higher the perch, the farther away the target and the more difficult it is to see. A lower limit is set by the reduced amount of water surface visible from low perches — the lower the perch, the less water visible and the lower the probability of detecting a fish of suitable size in a given time interval. These two conflicting variables create a situation for optimization, with each kingfisher potentially selecting the perch height that exposes it to a maximum amount of fish biomass per hour of search-time without making the fish targets too small. By selecting different perch heights, different-sized kingfishers may maximize their surveillance of prey of appropriate size (MacArthur 1972) and thereby maximize their capture rate as measured in units such as grams of fish per hour (minus energy spent in diving). Changes in surface fish density would affect optimum values, as is suggested by the significant changes in perch height at Salsipuedes. Local variability in perch availability is another variable to be considered. Certain perches with unobstructed views of the water, such as snags and leafless trees, may be selected regardless of perch height.

Some data suggest that perch height influenced prey size. For the two species (*amazona* and *americana*) for which the most data were available in the five-species community, prey length was correlated with height of perch from which the bird dove to obtain that prey (for *amazona*, r = .418, P < .001; for *americana*, r = .375, P < .001). In the

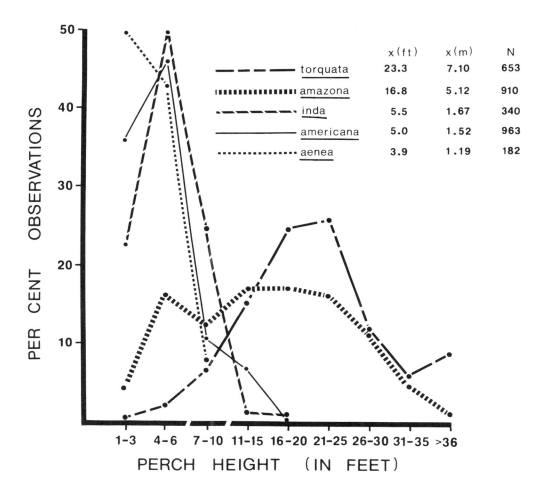

FIG. 10. Perch heights in five-species kingfisher community.

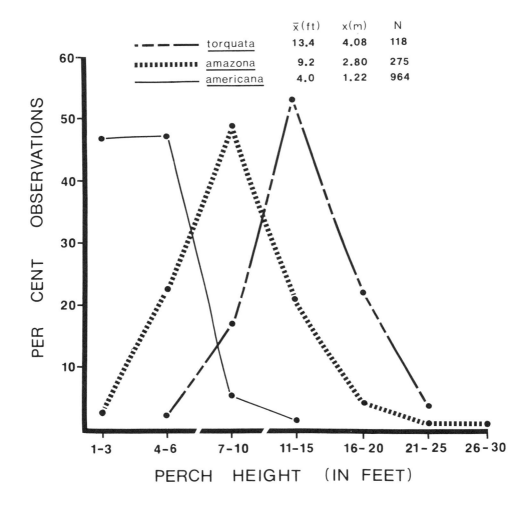

FIG. 11. Perch heights in the three-species kingfisher community.

three-species system, the same trend is evident for *americana* (r = .418, P = .014) but not for *amazona* (r = .101, P = .453). The correlation coefficients for the other species are not significant, but this may be caused by inadequate sample sizes; all are positive, with levels of significance ranging from .121 to .497.

HORIZONTAL LOCATION OF PERCHES

"Effective" shoreline was defined as that point at which the water surface ends and either land or the floating vegetation mat begins. In most situations, effective shoreline was delimited by the floating vegetation mat rather than the true shoreline itself, because kingfishers dive only into open water, not into floating vegetation, even though fish may be visible at the surface in the interstices of the mat. The "horizontal location" of a kingfisher hunting perch was defined as the distance from a point on the substrate directly beneath the perch to effective shoreline (rather than the diagonal from the perch directly to effective shoreline). These distances were either estimated to the nearest foot or measured using measure marks on the canoe. The distance was given a positive value for perches above water or a negative value for perches back from effective shoreline, i.e., above land or aquatic vegetation mats.

Comparing distributions of horizontal locations of hunting perches between species in the same community and within the same species in the two different communities (Table 13) yields highly significant differences between all species pairs (Kolmogorov-Smirnov test, P < .001, except *americana* vs. *inda* in the five-species system, P < .05), although the means differ by only a few inches in several cases. All species concentrated on the intervals -4 to +7 ft. from effective shore. Except for *torquata*, the mode for all species was in the 0-3 ft. interval. Because surface fishes are most abundant within the first 6 ft. from effective shoreline, concentration on this interval was expected.

The presence of *aenea* and *inda* in intervals farthest from shore is somewhat misleading because of differences in habitat structure in shaded versus open habitat. Most cases of perching far from effective shoreline in these two species were in flooded forest habitat, where water depth and abundance of surface fishes did not change with respect to distance from effective shore. Presence in these distance-from-shore intervals in flooded forest was not equivalent to presence in intervals far from shore in open-water habitat, where water was deeper and surface fishes fewer with increasing distance from shore.

Because not as much water is visible from a perch above land as is visible from one above water, all else being equal, perching above land reduces prey availability. Therefore, ideally no perches should be above land (or floating vegetation mats). Yet substantial proportions of perches of all five species had negative values for horizontal location, i.e., were not located above water. Again this reflects the limitations on perch selection imposed by habitat structure. In many cases, a given interval of shoreline could be attacked only from perches above land or floating vegetation mats rather than above open water. This was particularly pronounced at sites with extensive floating vegetation mats, which pushed effective shoreline farther from shoreline perches as the width of the mat increased. But the reduced area of water visible from perches above the mat was to some extent compensated for by the increased densities of fishes in areas with extensive floating vegetation mats (see fish density sections above).

TABLE 13. Distance of kingfisher hunting perches from shoreline. Data are presented as percentage of total number of observations for that species.

	Horizontal Location							
	-10 ft. or more	-9 to -5 ft.	-4 to -1 ft.	0 to +3 ft.	+4 to +7 ft.	+8 to +14 ft.	+15 ft. or more	N
Five-species community								
torquata	36.9	14.5	13.1	15.2	11.0	4.5	4.8	643
amazona	15.7	9.0	18.0	29.6	12.2	10.4	5.1	900
inda	—	1.3	20.4	63.1	12.7	1.3	1.3	314
americana	0.9	3.2	23.2	52.0	8.8	9.7	2.2	928
aenea	—	—	13.0	50.8	24.9	3.2	8.1	185
Three-species community								
torquata	—	16.1	42.4	32.2	9.3	—	—	118
amazona	0.4	1.5	24.1	59.8	14.2	—	—	274
americana	—	0.2	24.4	69.2	5.0	0.4	0.8	964

For a given distance back from effective shore, some low perches might be totally unacceptable, providing too small an area of water visible from that perch or forcing the bird to enter the water at too low an angle. Even when overlooking open water, kingfishers did not enter water at angles of less than 11-27°, depending on species and study site (Remsen, in prep.). Thus, there may be some minimum acceptable height for perches not overlooking open water, and this height should increase as distance back from effective shoreline increases. To examine this, perch height is plotted against horizontal location of perches (Figs. 12-14) and regression lines are plotted, first for perches not above open water and then for perches above water, for three species, *torquata*, *amazona*, and *americana*, in the five-species community. In all three cases, there is a highly significant ($P < .00001$) negative correlation between perch height and horizontal location for those perches back from effective shoreline. This same relationship holds for these three species in the three-species community, but breaks down for *inda* and *aenea*. For *inda*, the slope (-.121)

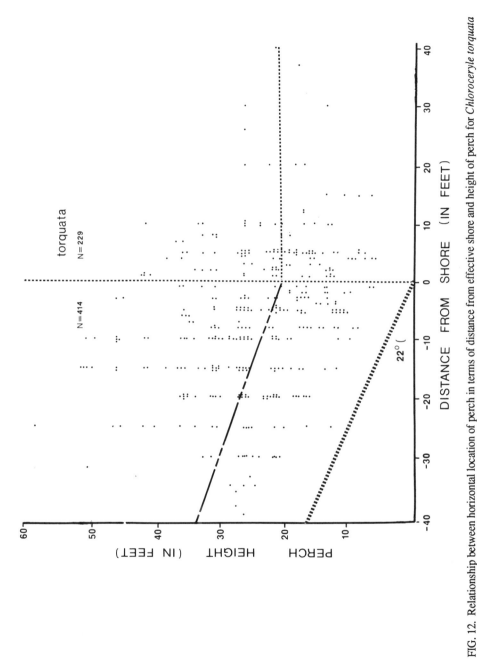

FIG. 12. Relationship between horizontal location of perch in terms of distance from effective shore and height of perch for *Chloroceryle torquata* in the five-species community. The regression line for perches back from the shoreline is $y = .305x + 20.9$, with $r = -.282$ and $P < .00001$. For perches over water, the regression line is $y = .021x + 20.8$, with $r = .019$; this line does not differ significantly ($P = .772$) from a straight line with slope $= 0$.

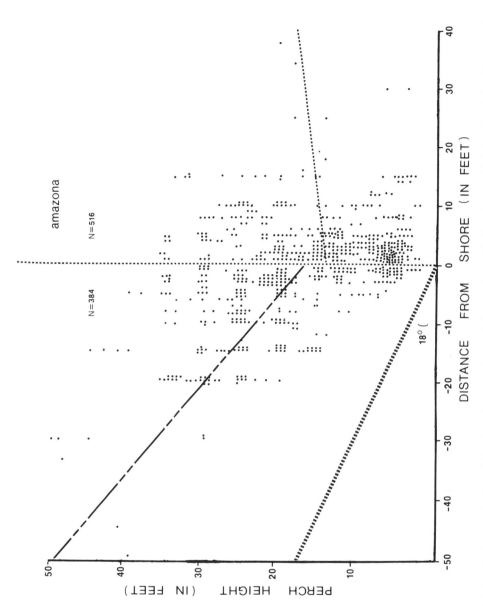

FIG. 13. Relationship between horizontal location of perch in terms of distance from effective shore and height of perch for *Chloroceryle amazona* in the five-species community. The regression line for perches back from the shoreline is y = -.674x + 16.0, with r = -.532 and P < .00001. For perches over water, the regression line is y = .108x + 12.7, with r = .088; this line does not differ significantly (P = .146) from a straight line with slope = 0.

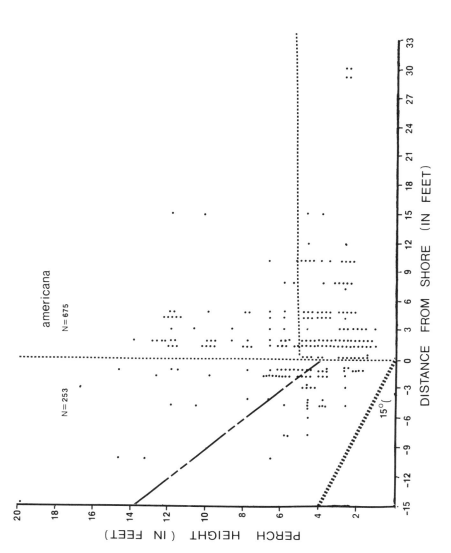

FIG. 14. Relationship between horizontal location of perch in terms of distance from effective shore and height of perch for *Chloroceryle americana* in the five-species community. Each point represents two observations. The regression line for perches back from the shoreline is y = -.649x + 4.1, with r = -.474 and P < .00001. For perches over water, the regression line is y = .006x + 4.8, with r = .010; this line does not differ significantly (P = .789) from a straight line with slope = 0.

is in the expected direction despite lack of significance (P = .442, r = .095). For *aenea*, however, the slope (1.24) was counter to the trend for the other species (P = .008, r = .528). This discrepancy for *aenea* is probably due to the lack of any perches farther back than -3 ft. from effective shoreline, where the effects of not being directly above open water are minimal. For all species, when only the perches above water are considered, no significant relationship (P > .05) is found between perch height and horizontal location.

There seems to be some minimal acceptable perch height for perches not above water that increases with increasing distance away from shore. In Figures 12-14, one can easily visualize a boundary line, paralleling the regression lines and forming a lower limit for perch heights, that increases with increasing distance from effective shore. Plotted on all three graphs is the line for the minimum angle of dive entry, empirically determined for each species (22° for *torquata*, 18° for *amazona*, and 15° for *americana*) when perched over open water; it seems that these boundary lines have steeper slopes than the lines for these observed minimum dive angles.

DIVE ENTRY POINT

For each dive entry point, the distance to nearest effective shoreline was estimated to the nearest foot. This could often be measured precisely, because ripples radiating from the dive entry point persisted for up to a minute after the dive, and measure marks on the canoe could be used to measure the distance to shore.

The three species that frequented open habitats (*torquata*, *amazona*, *americana*) showed a slight tendency to dive farther from shore than the two species of shaded habitats (*inda* and *aenea*); but in general, all five species concentrated on the first 6 ft. from shore, with by far the great majority of dives in the first 3 ft. (Table 14). The only species significantly different from one another in the five-species community were *amazona* vs. *americana* and *amazona* vs. *aenea* (P < .05, Kolmogorov-Smirnov test). In the three-species community, dive entry points were even more compressed into the first 3 ft. from shore. Part of the reason for the scarcity of entry points farther than 10 ft. from shore in the three-species community at Salsipuedes was that in only a few sections was the stream more than 20 ft. in width; but this does not explain the reduction in frequency of dives in the 4-6 and 7-9 ft. intervals, a reduction statistically significant at least for *americana* (P < .001).

All kingfishers in both communities made most of their dives in the 0-3 ft. interval. I develop below three separate arguments to show that kingfishers actively select the 0-3 ft. interval.

Using mean perch height, mean horizontal location of perch, and minimum angle of dive entry into the water, empirically determined for each species (Remsen, in prep.), hypothetical maximum distances from shore for dive entry points can be calculated. In the five-species system, these are 52.9 ft. for *torquata*, 51.6 ft. for *amazona*, 21.9 ft. for *inda*, 20.4 ft. for *americana*, and 24.0 ft. for *aenea*. Looking at maximum distances recorded for dive entries for each species (20, 20, 10, 15, and 5 ft., respectively; see Table 14), it can be seen that the potential maximum distances were never realized. Another way of looking at this is to compare the proportion of long (more than twice the height of the perch) lateral

TABLE 14. Distances from shoreline for kingfisher dive entry points. Data are presented as percentage of total number of observations.

	Dive Entry Point						
	0-3 ft	4-6 ft.	7-9 ft.	10+ ft.	mean	max	N
Five-species community							
torquata	72.1	16.4	—	11.5	2.9	20	61
amazona	69.4	15.3	7.1	8.2	3.3	20	196
inda	83.9	13.8	2.3	—	1.8	10	87
americana	83.0	8.5	2.1	6.4	2.3	15	282
aenea	90.3	9.7	—	—	1.4	5	62
Three-species community							
torquata	91.4	2.9	5.7	—	1.6	8	35
amazona	84.8	15.2	—	—	2.0	5	92
americana	98.6	1.4	—	—	1.1	5	418

distances of dive entries from perch (not from shore) with the proportion of dive entries near and far from shore. The proportion of long lateral distances is significantly higher for dive entries within the first 3 ft. from shore (Chi square, $P < .01$) for all species except *aenea* in both communities. Thus, when the dive entry point was a relatively long distance from the perch, these entry points were close to effective shore; in other words, the birds' dive paths when diving long distances were more or less parallel to shore rather than perpendicular.

Another method for showing that kingfishers actively select the first 3 ft. from shore is to take the mean perch height, mean horizontal perch location, and mean dive angle for each species to calculate a mean lateral distance, d, from perch to dive entry point. This distance is then used to form a circle with radius d and center at the hunting perch. That portion of the circumference that crosses water, c, will be larger than that of a semicircle for perches recessed from the shoreline. If kingfishers dove randomly with respect to distance of dive entry point from shore, then dive entry points should be distributed randomly along the above-water perimeter, c, rather than clustered in that portion of c within 3 ft. of shore, c_X; the proportion of dives along that portion of c for which the distance from shore was less than 3 ft. would have been equal to $2c_X/c$, and the proportion of dives along c for which distances from shore was greater than 3 ft. would have been $c-(2c_X/c)$. Because none of the kingfishers' mean horizontal perch locations was exactly at "0," the calculations become complex. Taking this into account, the proportion of dives that should have occurred within 3 ft. from shore, if kingfishers were diving randomly with respect to distance from shore, are .10 for *torquata*, .19 for *amazona*, and .39 for *americana* in the five-species situation, and in the three-species situation, .24, .29, and .38, respectively. The observed proportions were, in the five-species community, .72, .69, and .83, and in the three-species community, .91, .85, and .99, respectively (Table 14). The differences between expected and observed for each species are highly significant (Chi square, $P < .0001$). These calculations were not performed for *inda* or *aenea*, because over 90% of their perch locations were in shaded habitat or recessed pools, which were structurally much more complex: a kingfisher could dive at its maximum distance away from the perch and have a high probability of entering the water less than 3 ft. from some opposite shore.

A third way to look at the clustering of dive entry points close to shore is to look at their observed mean distances from shore (Table 14) in comparison to the expected mean distances if the kingfishers were diving randomly with respect to distance from shore of dive entry. Using mean perch height, mean horizontal perch location, and mean dive angle, these expected means in the five-species community are 11.2 ft. for *torquata*, 6.9 ft. for *amazona*, and 3.8 ft. for *americana*, vs. observed means of 2.9, 3.3, and 2.8 ft., respectively. In the three-species community, the expected means are 5.3, 4.0, and 3.0 ft. vs. observed means of 1.6, 2.0, and 1.1 ft., respectively. Thus all kingfishers in open habitats are diving much closer to shore than expected by chance. In this, and the previous calculations for proportions of dives within 3 ft., use of means for perch height, dive angle, and horizontal perch locations assumes a normal, or at least non-skewed, distribution for these variables, which is not the case for most species. As a first-order approximation, however, the evidence seems convincing that kingfishers actively select the first 3 ft. from shore for dive entry points.

Why concentrate on the first 3 ft. from shore? Surface fish densities declined substantially beyond the first 6 ft. from shore, as noted earlier (Fish Data), and if differences in fish densities between the 0-3 and 3-6 ft. intervals had been quantified, my qualitative but extensive observations during the fish censuses indicated that the differences would certainly have been significant: the closer to shoreline floating vegetation, the greater the density of surface fishes. Thus, the kingfishers don't really have much opportunity to segregate on the basis of distance from shore of dive entry point, because almost all surface fishes are within the first 6 ft., and especially the first 3 ft.

PREY SIZE

The standard length (tail-fin not included) of each fish was estimated as a ratio of the kingfisher's bill length while the fish was being manipulated in the bill before swallowing; Willard (1985) also used this technique to quantify size of prey taken by kingfishers. The mean bill length of the Amazon Basin populations of each kingfisher (Table 1) was then used to calculate fish length in millimeters. I spent considerable time practicing the judging of ratios of prey length to bill length, using paper models of kingfishers and fish of various sizes at various distances. In practice sessions, my estimated prey length came within 10%

of actual length for items less than twice the length of the bill, and within 20% for larger prey items. Small fish in small kingfisher bills were more difficult to see, but this was compensated for by the closer approach tolerated by the small kingfishers.

Although prey-length to bill-length ratios have been used frequently to estimate prey size in studies of fish-eating birds, the accuracy of such methods has been questioned (Bayer 1985, Goss-Custard et al., 1987). Although most such criticisms of the method do not apply to kingfishers, from the outset, I regarded this method as an unsatisfying means of measuring such a critical parameter. Nonetheless, there was no feasible alternative. Collecting birds from limited populations seemed unwise. Many individual birds were invaluable in that they had grown accustomed to my presence, allowing close approach; "retraining" new birds would have been costly in terms of time. A few birds were collected away from the intensive sites for stomach contents, but these yielded only masses of fish scales. At the conclusion of the study, four individuals on the intensive sites were shot immediately after a prey capture; my estimates of prey length before collecting the birds were within 10% of actual length in all four cases.

Another estimate of prey size was also recorded: the "depth" of the fish at its deepest point. This was estimated as a ratio of prey depth to bill depth (at the base of the culmen). Accuracy of this estimate was much lower than for prey length for several reasons (Remsen 1978), and because the outcomes of analyses incorporating prey depth as an additional measure of prey size did not differ from those using prey length only (Remsen 1978), these additional analyses are not included here. Fortunately, fishes taken by kingfishers were generally similar in shape (fusiform), except that *torquata* took a higher proportion of deep-bodied (compressiform) fishes. Thus the prey size estimates for *torquata* based on length only are biased towards small size.

Frequency distribution for lengths of fish prey for each kingfisher species in the five- and three-species communities (Fig. 15) are all significantly different from one another (Kolmogorov-Smirnov test, ($P < .05$) except for *inda* vs. *americana* in the five-species community, although this difference is also significant using the Mann-Whitney U test ($P < .01$). The prey-length distributions for *amazona* and *americana* shifted significantly downward toward smaller fishes from the five- to three-species site (Kolmogorov-Smirnov test, $P < .01$), but there was no change for *torquata*. Mean prey lengths at the five-species sites were 29.6 mm for *aenea*, 41.1 mm for *americana*, 47.0 mm for *inda*, 83.1 mm for *amazona*, and 92.6 mm for *torquata*. These were generally similar to those found for the five species at a site in Amazonian Peru by Willard (1985), especially considering that the data sets were produced by visual estimates by two different observers. At the three-species site, mean prey lengths were 19.4 mm for *americana*, 42.8 mm for *amazona*, and 91.0 mm for *torquata*.

Therefore, in these five species of kingfishers, although some ecological separation with respect to habitat differences exists, the primary axis of separation is differences in prey size. That Neotropical kingfishers "obey" the "competitive exclusion principle" is really a trivial result, as argued by Hardin (1960), Slobodkin (1962), and Schoener (1974). The purpose of this study was not to produce yet another "verification" of the competitive exclusion principle, but to test hypotheses concerning mechanisms of species-packing.

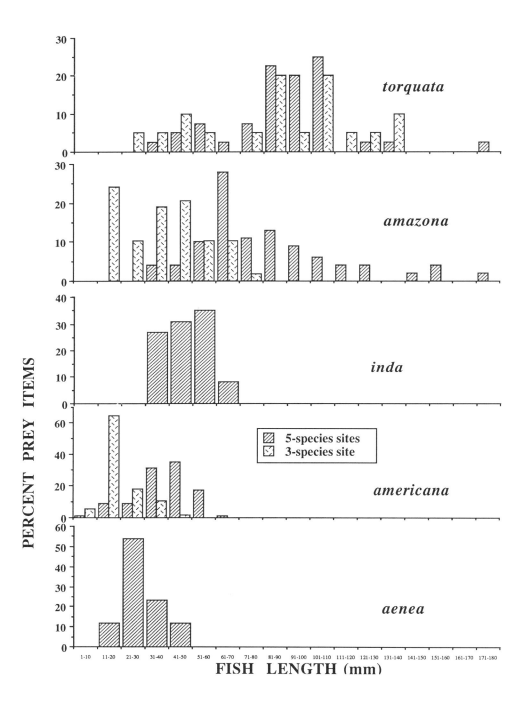

FIG. 15. Lengths of kingfisher prey in the five- and three-species communities.

Lack (1971) has provided abundant evidence that closely related birds are ecologically separated. As Schoener (1974) so clearly stated, further research, the sole purpose of which is to document more instances of resource partitioning, is of minor value, other than providing quantitative natural-history data on another series of potentially competing species.

In these five species of kingfishers, mean length of prey was strongly correlated with mean bill length, bill width, or bill depth (Spearman rank, $r = 1.00$, $P < .01$). Such among-species correlations between prey size and predator body size or bill size have been found in several series of closely related bird species that are potential competitors (Hespenheide [1973] and references therein), including many sympatric piscivorous birds (Dorward 1962; Ashmole 1968; Harris 1970; Lemmetyinen 1973; Din and Eltringham 1974; Willard 1977; Baltz et al. 1979; Whitfield and Blaber 1979; Hulsman 1981, 1987; Knopf and Kennedy 1981; Hom 1983; Diamond 1984), although the relationship does not always hold over a small range of body sizes (Diamond 1984). In the kingfishers, it is critical to note that the prey size taken by two species, *americana* and *amazona*, shifts dramatically between the five- and three-species sites, but their relative position on the prey-size:body-size gradient does not change. Whether such correlations between prey size and body size are responses to interspecific competition is, of course, a separate question that, for kingfishers, will be addressed elsewhere (Remsen, in prep.).

In general, the more similar two kingfisher species are in bill length, the more similar they are in their prey length. In the five-species community, mean lengths of prey taken by *torquata* and *amazona* are more similar than those for any other species pair. These are the two species closest in bill length (Table 1). The pair with the second-closest bill lengths, *inda* and *americana*, showed the second-greatest similarity in mean prey length. Similarly, adjacent species on the size continuum with the greatest disparity in bill length, *aenea* vs. *americana* and *inda* vs. *amazona*, are those pairs with the greatest differences in prey lengths, but the rank-correlation for degree of difference in prey length and bill length is not perfect: *aenea* and *americana* have the greatest bill-length ratio, but *amazona* and *inda* have the greatest mean-prey-length ratio. Nevertheless, the close agreement is interesting, lending some support to the validity of indices of morphological similarity as measures of ecological similarity (Cody 1974, Ricklefs and Cox 1977, Ricklefs and Travis 1980). Note that it is possible for the same prey-length vs. bill-length correlation to hold with no correspondence, or even with reversed correspondence, between degree of difference in the prey length and bill length. The probability that such close correspondence (only one rank or less out of place) in degree of difference occurred by chance is .166 (4/24).

What determines the frequency distributions of prey lengths for each species? Even if there were no potential competitors present, an upper limit to prey length would be set by limitations on the size of fish that can be retrieved from the water and swallowed. Kingfishers may be able to catch fish larger than they can swallow. David Willard (personal communication) has witnessed such an instance for *torquata*. For each species, a lower limit to prey size is set by what can be physically seen or captured by diving and, assuming time and energy are limited, by what is worthwhile in terms of energy investment.

Within these limits, kingfishers should select as large a fish as possible. The larger the fish, the fewer dives needed to obtain an equivalent amount of energy.

Minimizing the number of dives would be important for several reasons. Dives themselves are costly events in terms of energy expenditure relative to other kingfisher activities. Also, the amount of time spent preening is probably proportional to the number of dives. Furthermore, each time a kingfisher dives, it makes itself more conspicuous to potential predators above water, such as hawks, and those below water, such as large fishes. Finally, every dive makes the fishes in the bird's territory potentially more alert to predation from above. Factors that would tend to increase the number of dives include the possibility that small fishes are easier to catch than large fishes, because larger fish may be more wary; on the other hand, smaller fish make smaller targets. Also, because large fish are scarce, a kingfisher may not be able to afford to wait for the largest fish it can catch. Unfortunately, difficulties in measurement of most of the above parameters will impede any analysis of "optimal foraging" in kingfishers. Studies of other piscivorous birds have found that, within the range of sizes available and of sizes that can be handled successfully by the birds, the largest fishes are selected (Britton and Moser 1982, Doornbos 1984).

PREY TYPE

The percentages of prey items in various unequal taxonomic categories, corresponding to my visual discriminatory abilities, for each species in the five-species community are given in Table 15. Data for the three-species community are not presented quantitatively because, with a much lower absolute number of fish observed in the censuses, diversity of taxonomic categories was low. Also, because most prey items were small relative to the bills of *amazona* and *americana* (Fig. 15) and were eaten so quickly because of this small size, few could be identified with confidence (see also Cezilly and Wallace 1988). Qualitatively, all three species at Salsipuedes were taking primarily Characidae, supplemented by substantial percentages (5-15%) of Loricariidae. A crab taken by *torquata* on one occasion was the only non-fish prey item observed at Salsipuedes.

Taxonomic composition of the diet seemed to reflect what was available on the surface, although statistical analysis of such small sample sizes for prey items, coupled with low percentages of availability for most categories, was difficult.

1. *Characidae*: The proportion of Characidae in each species' diet was not statistically different from the expected frequency calculated from the percentage of Characidae in the surface fish population for the size-classes of fish in the kingfisher diet (Chi square, $P >$ .05). Whether kingfishers were selecting or avoiding certain species or genera of characids was beyond the scope of this study. Innovative methods such as those used by Tjomlid (1973), Lemmetyinen (1973), Douthwaite (1976), Doornbos (1979, 1984), Ainley et al. (1981), Pilon et al. (1983), and Gales (1988) to identify fishes from regurgitated pellets, partially digested remains in the stomach, or otoliths depend on a simple fish fauna with few prey species, and would be extremely tedious and perhaps impossible with the extremely diverse fish fauna of the Amazon Basin.

It is doubtful that kingfishers make fine distinctions between fish species or genera. Cloudiness and darkness of the water, especially during high-water season, probably decreases the possibility for perceiving such distinctions. There is no clear reason why a kingfisher should ignore any prey item small enough to capture yet large enough to be energetically worthwhile, unless distasteful, difficult to swallow, or disproportionately difficult to catch. Ashmole (1968) presented similar arguments concerning prey selection in fish-eating terns. Zottoli (1976), Whitfield and Blaber (1978, 1979), and Härkönen (1988) found that differences in taxonomic composition of prey of piscivorous birds were mainly the consequence of habitat differences in the birds and their fish prey rather than any tendency toward taxonomic selectivity. There is some evidence, however, that indicates that for at least one piscivorous bird, *Pandion haliaetus*, foraging success is directly influenced by prey type (Swenson 1979).

2. *Cichlasoma festivum:* This was one of the most common species of fishes on the surface (Fig. 5) and may be the single most important species in terms of biomass. This cichlid is unusual among the Cichlidae in that most members of the family are benthic or middle strata fishes (see discussion in Remsen 1978).

TABLE 15. Prey types taken by kingfishers at the five-species study sites. Categories marked with asterisks are those either selected or avoided by kingfishers (P< .01/12 = .0008).

	Percent of fish population	Percent of Diz				
		torquata	*amazona*	*inda*	*americana*	*aenea*
Characidae	80.4	80.8	81.1	77.4	88.0	96.3
Cichlasoma festivum	9.7	—	—*	—	—*	—
other Cichlidae	2.1	12.8	5.4	3.2	—	—
Cyprinodontidae	4.6	—	2.7	6.5	0.8	3.7
Anostomidae	2.2	—	—	—	—	—
Loricariidae	0.2	—	2.7	—	—	—
Gasteropelicidae	0.1	—	—	—	—	—
? *Hydrocinus*	0.1	4.3	8.1*	—	—	—
Electrophoridae	0.1	—	—	—	—	—
Pimelodidae	0.1	—	—	—	—	—
shrimp	0.5	—	—	12.9*	11.2*	—
crab	0.1	2.1	—	—	—	—
N	6923	47	74	31	125	27

Despite its abundance on the surface (up to 20% of all fish on the lakes of ISS; see Fig. 5), and a range of sizes placing it within the prey-size preferences of all five kingfishers, it was never seen to be captured by any kingfisher. This cichlid would have been one of the most easily identifiable fishes in the grasp of a kingfisher bill: its conspicuous diagonal stripe, conspicuously prolonged pelvic fins, and distinctive general shape make recognition easy. Because *C. festivum* changes color pattern when agitated, the diagonal body stripe may have been obscured when captured. Nevertheless, only a few captured cichlids were observed with the vertically barred "agitation" pattern, and these lacked prolonged pelvic fins. Thus I am reasonably confident that kingfisher predation on *C. festivum* was extremely rare, despite the abundance of this conspicuously marked fish on the surface.

C. festivum seems to be exceedingly adept at avoiding surface predation. In my attempts to dip-net samples of surface fishes for identification, this species proved virtually impossible to capture. Simple experiments with this species in an aquarium also indicated that it is extremely alert to surface predation, much more so than a wide range of characins. Perhaps the large eyes, always oriented toward the surface in this species' characteristic, upward-tilted posture, along with the superior learning abilities displayed by many cichlids, combine to give *C. festivum* a near immunity to kingfisher predation (see Remsen 1978 for fuller discussion).

Because *C. festivum* was not taken by kingfishers, densities of this fish could not really be included in calculations of kingfisher prey availability and were therefore removed from surface fish-density data with respect to kingfisher predation data (i.e., Fig. 7). This removal had the effect of bringing the fish density in lakes at low-water season, where *C. festivum* was extremely common, closer to stream densities, and bringing both values closer to those of completely shaded habitats, where *C. festivum* was rare, and to high-water densities in general, when this species was nearly absent from surface fish censuses.

3. *Other fishes*: The Poisson distribution was used to approximate the probability associated with numbers of prey items observed in the rarer fish categories. The frequency of a fish type in the fish censuses was used as the expected frequency of a rare event in the Poisson distribution. The probability that the observed frequency in each kingfisher's diet would occur by chance was then calculated using the Poisson series. Levels of significance were set conservatively because of the large number of tests performed. Only one category for one kingfisher species had an associated probability of less than .001, i.e., was definitely being selected from what was available: *amazona* was selecting a long, narrow characin that may have been *Hydrocinus* sp. or *Acestorhynchus* sp. (unfortunately, no voucher specimens could be obtained).

4. *Aquatic invertebrates*: Both *inda* and *americana* were selecting ($P < .01$) a shrimp (Palaemonoida, probably Palaemonidae; specimens deposited at California Academy of Sciences) only rarely seen clinging to floating vegetation in the visual censuses. Crabs (Trichodactylidae; specimens deposited at California Academy of Sciences) were occasionally taken by *torquata* at ISS and Salsipuedes. It is possible that *torquata* was the only species large enough to handle and swallow these crabs. Apparently the congeneric *Ceryle maxima* of Africa sometimes specializes locally on freshwater crabs (Forshaw 1983).

Several primarily fish-eating kingfishers of the Old World have been reported occasionally to eat organisms other than fishes or aquatic invertebrates (Forshaw 1983 and references therein). *Ceryle alcyon* of North America has also been reported taking a great diversity of prey, including crayfish, mayflies, salamanders, frogs, and mice (Davis 1982; references in Forshaw 1983). No such diversity has yet been reported in Neotropical species. Studies so far have indicated that the five Neotropical species eat aquatic animals almost exclusively, primarily fishes with some invertebrates (Skutch 1957, 1972; Willard 1985; this study), with non-aquatic prey taken extremely rarely. For example, Greene et al. (1978) saw an unsuccessful attempt by *torquata* to catch juvenile iguanas crossing water, and Willard (1985; personal communication) saw a *torquata* catch a dragonfly as it flew by the bird's perch.

One species, *aenea*, has often been reported to be insectivorous (Blake 1953, ffrench 1973, Ridgely 1976, Forshaw 1983). Meyer de Schauensee (1970:167) even stated that this species is entirely insectivorous. In many hours of watching *aenea*, including time in which it continued to fish while termite swarms attracted many other birds, I never saw it attempt to catch an insect.

In my opinion, the published reports of *aenea* "flycatching" (e.g., Dickey and van Rossem [1938]) require verification. When *aenea* and *americana* "pull out" of a dive before entering the water (when the target fish seemingly disappears while the kingfisher is in midair), they abruptly reverse direction in flight, returning to the same perch. The immediate impression is that the bird has sallied out for a flying insect, flycatcher fashion. I have seen an experienced observer assume that an *americana* had sallied out for an insect after such a "pull-out" dive, and I feel that this type of behavior may be responsible for some of the statements in the literature that sallying for insects has been observed in *aenea*. Also, perhaps the tendency for *aenea* to move from one stream to another overland (instead of strictly following the watercourse, as usually do other kingfishers), and its consequent appearance in forest mist-net samples, may lead to the assumption that it has been foraging away from water (James R. Karr, personal communication).

I have subsequently contacted many persons with extensive field experience in the lowland tropics (including John Fitzpatrick, James Karr, John O'Neill, Ted Parker, Robert Ridgely, Alexander Skutch, Paul Slud, Dave Willard, and Edwin Willis), and none has ever seen *aenea* take anything but fishes. The only definite report that I could find of insectivory was that of Thomas R. Howell (personal communication), who collected an *aenea* in Nicaragua that had insect parts in its stomach. Of the 25 specimens with stomach contents data in the Louisiana State University Museum of Zoology collection, only one contained any insect remains: a beetle elytra among many fish scales. Small fishes were the only items found in stomachs of birds collected in Surinam by Haverschmidt (1968). I conclude that although *aenea* may occasionally take some insects, the degree to which it is insectivorous has been greatly exaggerated in the literature.

OTHER PISCIVOROUS BIRDS

Kingfishers were not the only organisms, of course, that ate fishes on the study sites. With evidence from other studies of competition or heavy overlap between some birds and other piscivorous vertebrates for prey (Eriksson 1979, Eadie and Keast 1982, Härkönen 1988), competition between kingfishers and fish is possible, but measurement of such was beyond the scope of this study. Terns, herons, hawks, and others also contributed to the piscivorous avifauna (Tables 8-10), although none was as common as kingfishers. To investigate the possibility that these other piscivorous birds played a role in kingfisher community interactions, an attempt was made to quantify foraging behavior of the common species. Unfortunately, sample sizes large enough to calculate overlap values with kingfishers could not be obtained. Even on a qualitative basis, however, most other piscivorous birds differed substantially from kingfisher in prey size, prey type, or feeding site selection, thereby minimizing ecological overlap. Data and natural history observations are presented below for the most important species (and the remaining species are discussed in Remsen 1978).

OSPREY (*Pandion haliaetus*)

One or two *Pandion haliaetus* were resident on the lakes of ISS even through the summer months, when most of the world's population breeds in the Northern Hemisphere. At least two, and perhaps three, individuals foraged over Lake Tumi Chucua. *P. haliaetus* was rarely seen at the stream study sites, Quebrada Tucuchira and Arroyo Salsipuedes. In general, this species foraged over lakes or extensive shallow shoals on large rivers and was rarely seen elsewhere.

P. haliaetus hunted by cruising in a straight line over long expanses of open water. Occasionally, *P. haliaetus* perched in shoreline trees overlooking water were suspected of hunting from those perches, but they were never seen to launch from a perch directly to a capture attempt. Fish were usually caught by the bird's dropping gradually to the water and grabbing the target with its feet. In contrast to the nearly vertical dives of terns, the final approaches of *P. haliaetus* to prey were at virtually parallel planes to the water surface.

The mean estimated cruising height of hunting *P. haliaetus* was 61.3 ft. (range 20-120 ft., N = 114). This differs somewhat from the mean cruising height, 42.6 ft. (N = 13), found by Willard (1985) for this species on an oxbow lake in Amazonian Peru. The mean estimated distance from effective shoreline was 277.3 ft. (range 5-1000 ft., N = 114). One of the prey items seemed to be a red-bellied piranha (*Serrasalmus* sp.), but the rest, all fishes, were unidentifiable. Only five were seen well enough to estimate their lengths; their mean length was 165 mm (range 127-203 mm), and mean depth was 76.2 mm (range 51-102 mm). This is considerably smaller than the mean fish length taken by *P. haliaetus* in Wyoming (Swenson 1978), Finland (Häkkinen 1978), and Amazonian Peru (Willard 1985), but is nearly identical to the mean weighted prey length (162.7 mm) in Minnesota (calculated from Dunstan 1974).

Sizes of fishes taken by *P. haliaetus* at various localities probably reflect the size of whatever fishes happen to be common at the surface, rather than competitive interactions with other piscivorous birds. *P. haliaetus* was capable of taking much larger fishes than any other piscivorous bird except some of the larger, shore-bound herons, and the majority of its hunting time was spent far from shore, where it hunted for fishes unavailable to shoreline piscivores. Only *Phaetusa simplex* (see below) regularly foraged at great distances from shore, and it overlapped only slightly with *P. haliaetus* in size of prey taken. *P. haliaetus* overlapped little with kingfishers; it took fishes too large for any but *amazona* and *torquata* and foraged at much greater distances from shore.

LARGE-BILLED TERN (*Phaetusa simplex*)

Phaetusa simplex was found on rivers as large as the Amazon to those as small as 20-50 m wide, such as the lower Río Cayarú Observations were few from narrower rivers such as Quebrada Tucuchira (Table 10), where they were seen only at the widest portion, section C. Apparently the foraging of this species is impeded in narrow channels by encroachment by the canopy on open space above the water. Away from nesting sites, *P. simplex* was usually seen alone or in twos and small flocks. Densities on one section of the Amazon between ISS and Leticia averaged about 0.4 bird/km most of the year. Occasionally, large concentrations were encountered, such as the 500 estimated in 15 km along the lower Río Cayarú and an adjacent narrow channel of the Amazon on 6 July 1974.

P. simplex was found regularly on the lakes at ISS, averaging 1.4 birds per census (Table 9). As many as 25 individuals were seen on a single day there, but never more than 3 on census days. In contrast to data from adjacent rivers, densities at high-water and low-water did not differ. Strangely, this species was not recorded at Lake Tumi Chucua, although known to be present on the Río Beni nearby.

P. simplex foraged by cruising in a more or less straight line over open water, plunging nearly vertically into the water when attempting prey capture, in typical tern fashion. On one occasion, a bird was seen to skim the water briefly (4-6 m), in a manner typical of skimmers (Rynchopidae); this was the only observation of this type of behavior in many hours of observation. Willard (1985) also reported two such instances of skimming.

The mean estimated hunting height (Fig. 16) on ISS was 26.5 ft., close to the 33.8 ft. found by Willard (1985), and the mean height from which dives were initiated was 22.7 ft.; there was no significant difference between hunting heights from which dives were and were not initiated (Mann-Whitney U-test, $P > .05$). Successful dives were initiated from a mean height of 19.2 ft., vs. 25.0 ft. for unsuccessful dives, but this difference was not significant (Mann-Whitney U test, $P > .05$). These figures include data from both high-water and low-water seasons, which were not significantly different (Mann-Whitney U-test, $P > .05$).

The mean distance from effective shore for foraging birds (Fig. 17) was 53.1 ft., and this did not differ significantly (Mann-Whitney U-test, $P > .05$) from the mean distance from shore for dive entry points, 71.6 ft. Mean distance from shore for successful and unsuccessful dives differed significantly (22.9 vs. 106.1 ft., Mann-Whitney U-test, $P < .05$). There were no statistically significant differences in any of these data between high-water and low-water seasons.

Only nine prey items (all fishes) were seen well enough to estimate size relative to bill length. Mean prey length was 80.5 mm (range 56-125 mm); a similar distribution of prey lengths was found for this species in Amazonian Peru by Willard (1985). Thus, *P. simplex* took fishes of the size taken regularly only by the two largest kingfishers, *torquata* and *amazona*, but took these on the average much farther from shore than did the kingfishers. Furthermore, it was quite likely that species composition of the diets differed strongly, because composition of the surface fish fauna near shoreline vegetation probably differed from that of the open water used by the terns.

On Oct. 23, 1974, four *P. simplex* were observed hawking insects over the Amazon River along the shoreline of ISS for 5 minutes about 30 minutes before dark. They captured an unidentifiable, slowly flying insect about once every 15 seconds. This was the only observation of insectivory, and insects were certainly only a minor diet component, as found for this species in Amazonian Peru by Willard (1985).

YELLOW-BILLED TERN (*Sterna superciliaris*)

Sterna superciliaris, a much smaller tern than *Phaetusa simplex* (45 vs. 220 g), was most common over extensive shallow-water shoals on large rivers such as the Amazon and Javarí, particularly around mouths of large streams emptying into these rivers. Smaller numbers also used large lakes such as Lake Tumi Chucua and Camungo Lake (absent from the latter during high-water season). This species was never recorded on small streams such as Quebrada Tucuchira (Table 9) and was much rarer than *P. simplex* on smaller rivers such as the Cayarú (Table 10). *S. superciliaris* was thus much more restricted in its habitat preferences than *P. simplex* and much less common.

The mean estimated hunting height (no dives included) for *S. superciliaris* in open-water situations was 23.1 ft.(Fig. 16), and in recessed pools, 18.1 ft. (N = 21). Although distribution of these heights did not differ significantly between the two habitats

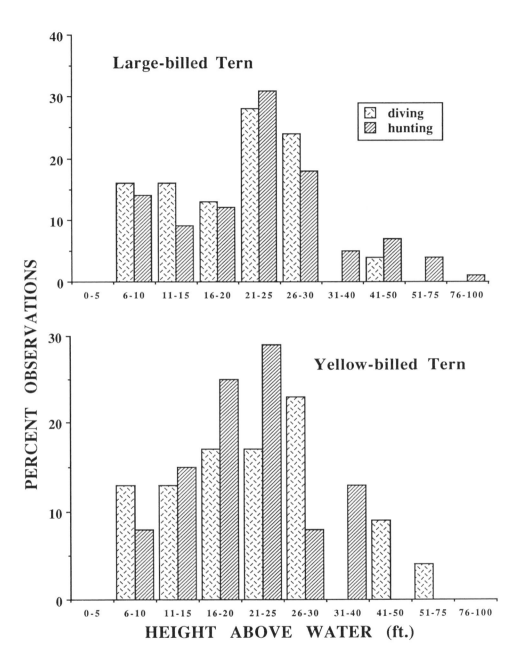

FIG. 16. Tern hunting heights.

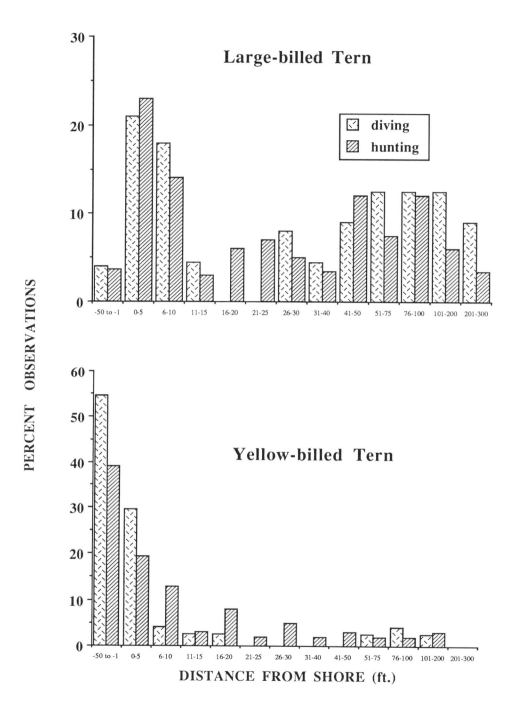

FIG. 17. Hunting distances from shore for terns.

(Kolmogorov-Smirnov test, P > .05), the distribution of heights of actual dives in the two habitats (mean 25.7 ft. in open water [N = 22], 14.3 ft. for recessed pools [N = 29]) was significantly different (Kolmogorov-Smirnov test, P < .01). No significant differences were found between dive heights from which successful vs. unsuccessful dives were initiated in either habitat (Median test, P > .05).

In contrast to the situation found by Willard (1985) in Amazonian Peru, where the mean cruising height of *S. superciliaris* was only 10 ft., the distribution of hunting heights (Fig. 17) for *S. superciliaris* did not differ significantly from open-water hunting heights of *P. simplex* (Kolmogorov-Smirnov test, P > .05). The distribution of dive heights also did not differ significantly between the two species (P > .05). Although the two terns did not differ in their hunting heights, there was a definite difference in habitat use. On ISS, *S. superciliaris* spent a large percentage of its time hunting over "recessed pools"; 46.6% of all observations were recorded at this habitat, compared to only 3.2% for *P. simplex*. Furthermore, when using open water, *S. superciliaris* foraged closer to shore than *P. simplex* (Fig. 17; Kolmogorov-Smirnov test, P < .05). The mean distance from shore in open water for *S. superciliaris* was 20.4 ft., vs. 53.1 ft. for *P. simplex*. At Lake Tumi Chucua, where *P. simplex* was not observed, *S. superciliaris* foraged at a greater distance from shore (x = 242.9 ft., vs. 20.4 ft. at ISS).

S. superciliaris took smaller prey than *P. simplex*, as expected from differences in bill lengths, 32.4 vs. 62.4 mm (measurements from Blake [1977]), and as found by Willard (1985) in Amazonian Peru. Mean prey length, estimated against bill length, was 15.1 mm (range 7-32 mm, N = 24); this was much smaller than prey taken by this species in Amazonian Peru, where Willard (1985) found that fish prey were all in the 22-45 mm range. The tiny size of many prey of *S. superciliaris* indicated that it may have been taking crustaceans or insects rather than fishes, although all items large enough (greater than 15 mm) to identify were fishes; in fact, Willard (1985) found that the majority of this species' prey was insects. Because fishes smaller than 15 mm were extremely rare on the surface at my sites, I feel that most tiny prey items must have been arthropods.

With its preference for open habitats and small prey, the only kingfisher likely to overlap to any appreciable extent with the *S. superciliaris* was *americana*. Prey-size overlap with *torquata* and *amazona* was extremely small.

BLACK-COLLARED HAWK (*Busarellus nigricollis*)

Busarellus nigricollis was present in small numbers at all intensive sites, but was most common on the lakes at ISS (Table 8), where one pair of adults and at least one immature were resident. In general, this species was found on lakes, small rivers, and streams wide enough not to be completely shaded by the canopy of shoreline vegetation. The largest number recorded in a single day was five on a 20 km stretch of the Río Cayarú on 16 Aug. 1975.

B. nigricollis hunted from elevated shoreline perches and captures fishes and perhaps other aquatic organisms by grabbing them with its feet from the water surface. The mean estimated perch height was 10.5 ft. (range 3-25 ft., N = 45); these data differ somewhat from those of Willard (1985), who found that mean perch height was about 4.3 ft. (range 3.3-9.8 ft., N = 14). Hunting perches (see discussion in preceding chapter for definitions of perch categories) were primarily densely or sparsely foliated trees (31.8% and 29.6%, respectively); also used were leafy bushes (15.9%), snags (11.4%), leafless trees (6.8%), and hidden trees (4.6%).

All strikes observed (N = 9) were directed at effective shoreline, recessed pools, or interstices of the floating vegetation mat. Strike entry points all ranged from +1 ft. from effective shore to -3 ft. (mean 0.0 ft.). This tendency for entry points to be within the floating vegetation mat contrasted with kingfishers, which were never seen to dive there; their entry points were not amidst vegetation, but always into open water. Willard (1985) also found that *B. nigricollis* was found primarily in "coves choked with water weeds." When the water surface is clogged with obstacles, capture of prey with the feet rather than with the bill is probably more effective, or at least less dangerous. If some fish species were found only or primarily in the interstices, then some ecological separation between *B. nigricollis* and kingfishers would result on the basis of differences in strike entry points.

Four prey items (all fishes) seen well enough to estimate size averaged 159 mm in length (range 102-203 mm) and 70 mm in depth. Thus this hawk probably catches prey much larger than can be handled by any but the two largest kingfishers, *torquata* and *amazona*. This size class of prey is nearly identical to that taken by *Pandion haliaetus*, but the latter took fishes from open water, whereas *B. nigricollis* concentrated on the shoreline; so, if fish populations or species in the two habitats differed, then little overlap in prey occurred between these two hawks. *B. nigricollis* has been reported feeding on a 150 mm-long frog (Willard 1985; D. Willard, personal communication), rats, mice, lizards, snails, and large insects, in addition to fish (Young 1929), including a 400 mm long eel and a 250-300 mm catfish (Morales et al. 1981), and so evidently takes some prey types never used by kingfishers or *P. haliaetus*.

LESSER KISKADEE (*Pitangus lictor*)

Pitangus lictor, a tyrannid of approximately the same body length as *Chloroceryle americana*, was observed capturing fishes on a few occasions at Lake Tumi Chucua. *P. lictor* was common at all the intensive sites (except Arroyo Salsipuedes, where observed only a few times), but the only site where fish-eating was observed was at the mouth of a small stream emptying into Lake Tumi Chucua, where small surface fishes were exceptionally common. *P. lictor* was always found in association with water; I never recorded it away from the shoreline of streams or lakes, but never saw it either on large rivers, such as the Amazon or Javarí, or on streams small enough to be completely shaded by shoreline canopy.

P. lictor usually preyed on insects taken from floating vegetation or foliage near the water's edge by "sallying" or hovering. Sallies for aerial insects were uncommon. It also sallied to the water surface for prey, most of which seemed to be insects fallen into the water. Virtually all foraging activity was concentrated within the first 2 m above water or shore. Although picking prey items, presumably insects, from water surfaces is a major foraging mode for this species (Fitzpatrick 1980), I can find no previous reports of piscivory.

At Lake Tumi Chucua, prey plucked from the water included a number of small fishes (primarily a small, silvery characid) and one palaemonoid shrimp. It never dove into the water but deftly snatched the fish from the surface. Eight characids seen well enough to estimate size relative to bill length averaged 30.8 mm in length (range 19-49 mm), and the one shrimp was estimated to be 39 mm long. These were the same species and sizes of prey taken from the same pool by the three smallest kingfishers, and overlap seemed to be heavy in this isolated instance, particularly with *Chloroceryle aenea*.

Although the related Great Kiskadee, *P. sulphuratus*, has been found to eat fish occasionally (Fitzpatrick 1980), it was not observed to do so at any of the sites.

GREEN-BACKED HERON (*Butorides striatus*)

This small heron was the most common and widely distributed of all the herons (Tables 8-10). It was the most abundant heron on all the intensive sites except the lakes on ISS, where, at least during low-water season, it was slightly outnumbered by *Casmerodius albus* (0.64 vs. 0.50 birds/census). It outnumbered all other herons combined on small streams such as Quebrada Tucuchira and Arroyo Salsipuedes. Some phenomenal densities were noted away from the intensive sites, e.g., 40 on 11 July 1974, in approximately 5 km along Quebrada Socó, a small stream flowing into the lower Río Javarí from the Brazilian bank.

This species was the only piscivore more common at high-water season than at low-water: for example, populations at Quebrada Tucuchira were four times as great at high-water season, and at ISS, six-and-one-half times as much. This trend was not restricted to the immediate vicinity of the intensive sites: counts along an approximately 25 km section of the Río Cayarú (Table 10) were 67 on 6 July 1974, at the end of high-water season, 50 on 16 July 1975, and 15 on 17 Aug. 1975, just prior to the peak of low-water season. It was unknown why seasonal population trends of this species were counter to those of all other piscivores. *B. striatus*, with its small size and water-level perches, does hunt closer to the water surface than any other piscivore, and perhaps this gives it a disproportionate advantage in hunting in the murky water of high-water season. Where this species went during low-water season, when aquatic habitats were the most restricted in area and when most piscivores were concentrated because of this restriction, was a mystery.

B. striatus hunted primarily using "stand and wait" and "walk slowly" techniques, as defined by Meyerriecks (1960) and Kushlan (1976b). The small size and agility of this species allowed it to use a much wider variety of perches than did the larger herons. It frequented mats of floating vegetation, particularly *Paspalum* mats, too flimsy to support the body weight of other herons; it also used small branches not strong enough to support other herons. It also perched on nearly vertical branches, something never observed in other heron species. A tabulation of hunting-perch types on the lake sites yielded the following (N = 91): floating vegetation (54.9%), snags (21.6%), logs (17.7%), leafy bushes (2.0%), leafless bushes (2.0%), and hidden bushes (2.0%). On the stream study sites, the breakdown (N = 39) was floating vegetation (61.5%), logs (15.4%), snags (12.8%), leafy bushes (5.1%), leafless bushes (2.6%), and hidden bushes (2.6%). The water was too deep at any of the intensive sites to permit wading by this small species.

Only three prey items were seen well enough to estimate size relative to the bill. One was an 18 mm-long aquatic arthropod; the other two were fish, both diamond-shaped characids, probably *Hyphessobrycon* sp., 42 and 49 mm long. At an oxbow lake in Amazonian Peru, Willard (1975) found that this heron took fish up to 70-90 mm in length, but the greatest number were in the < 23 mm size class; most prey items, however, were probably arthropods. In the llanos of Venezuela, Morales et al. (1981) observed six prey items taken by this species, all characins 30-100 mm long.

GREAT EGRET (*Casmerodius albus*)

C. albus was found at all the intensive sites, but was common only on Lake Tumi Chucua and the lakes on ISS (Table 9). Preferred habitat seemed to be extensive floating mats of *Paspalum* around lake borders wherever dense enough to support its body weight. It was much less common along the shorelines of streams and rivers, disappearing completely on small, canopy-closed streams. The largest concentration recorded was a flock of 63 scattered around the shoreline of the largest lake on ISS on 21 July 1974.

The only hunting methods recorded were "stand and wait" and "walk slowly" (Meyerriecks 1960, Kushlan 1976a). Although it occasionally hunted from firmly anchored shoreline brush and snags, most perches (98.8%, N = 82) were the sturdier portions of the floating vegetation mat. Most hunting perches were not on the effective shoreline but back from the shoreline, where it concentrated on interstices of the mat and around edges of small recessed pools. The mean distance from effective shoreline was -3.2 ft. (range 0 to -15 ft., N = 82). Only two prey items were seen well enough to be identified, both cichlids, 50 mm and 80 mm long. The most frequently recorded prey items for this species in southeast Peru were fishes, especially cichlids (*Aequidens* sp.), 45-90 mm long (Willard 1985). The most frequently recorded prey items for this species at a site in the llanos of Venezuela were also fishes, mainly Characidae (*Hyphessobrycon* sp.) and cichlids (*Aequidens* sp.), with mean lengths 25-30 mm (Morales et al. 1981).

C. albus probably overlapped with the two largest kingfishers, *torquata* and *amazona*, in prey size preference, but in general took prey from a microhabitat not used by kingfishers. Overlap was probably greater with *Busarellus nigricollis*, *Butorides striatus*, and *Ardea cocoi*, all of which also hunted in the floating vegetation mat. Willard (1985) found considerable differences in prey size and prey type among the three herons.

SUMMARY

Inadequate sample sizes of prey items for non-kingfisher piscivores prevented the quantitative calculation of overlap values with kingfishers to determine the importance of potential competition between them. Only three species, however, potentially show substantial overlap with kingfishers: *Butorides striatus* and *Sterna superciliaris*, with *Chloroceryle americana* in some lakes; and perhaps *Pitangus lictor* with *Chloroceryle aenea* in some situations.

This absence of substantial ecological overlap, however, does not exclude the possibility that competition between species in different bird families on an evolutionary time-scale has not affected community structure of piscivorous birds. For instance, in the absence of kingfishers, perhaps *Phaetusa simplex* might forage closer to shore, perhaps *Busarellus nigricollis* might forage more in open water, and perhaps *Pitangus lictor* might feed more extensively on fish. It is, however, difficult to envision any further niche expansion on the part of kingfishers in the absence of other piscivores without modification of current kingfisher foraging behavior to facilitate feeding in the interstices of floating vegetation or foraging aerially at greater distances from shore.

For a more complete survey of an entire community of piscivorous birds in Amazonia, see Willard (1985). As noted by Willard, our conclusions regarding resource use are similar. The major difference in community composition is that Willard's oxbow lake had much higher densities of swimming fish-eaters (*Phalacrocorax brasilianus* and *Anhinga anhinga*) and of most herons than were found at my sites; whether this difference is related to intrinsic ecological differences between our sites, or to greater human disturbance at my sites in the Leticia and Riberalta regions, is not known.

SPECIES-PACKING MECHANISMS

NICHE-BREADTH MECHANISM

A simplified, schematic representation of the niche-breadth mechanism of increased species-packing, in which resource-use curves are given for five hypothetical species competing for one resource, is presented in Figure 18. Niche breadth, here represented by the distance occupied by each species on the horizontal axis, i.e., the range of resources used by each species, is merely a measure of specialization. The number of species in Figure 18 is reduced from five to three without any changes in niche metrics except increasing niche breadth; niche overlap and resource base (the total distance on the x-axis occupied by this group of species) remain constant.

Theoretical approaches predict reduced niche breadths in more diverse communities (Levins 1968, MacArthur 1972), i.e., lower degrees of specialization in the less diverse community. This could occur because as the abundance of resources decreases, species are forced to broaden the range of resources used; some species are then forced to leave the community because of increased competition for the remaining resources. Also, species could be forced to leave because of the absence of some critical resource, and this absence then allows the remaining species to expand into the vacated range of resources. Finally, the greater unpredictability of resources in more temperate communities might prevent specialization on a narrow range of resource types. In any of these cases, the niche-breadth mechanism predicts that greater niche-breadths will be found in the less diverse communities.

Several different measures of niche breadth were computed for various resources thought to be important to kingfishers (e.g., prey length, prey type, habitat, dive entry point, perch height, perch type), and the results were consistent: niche breadth either did not change between the three- and five-species systems, or, in most cases, the change was counter to that predicted by the niche-breadth hypothesis — niche breadth usually decreased in the less diverse community. My data on niche breadths provide no evidence for the existence of competitive effects in kingfishers or for greater resource specialization in more diverse communities.

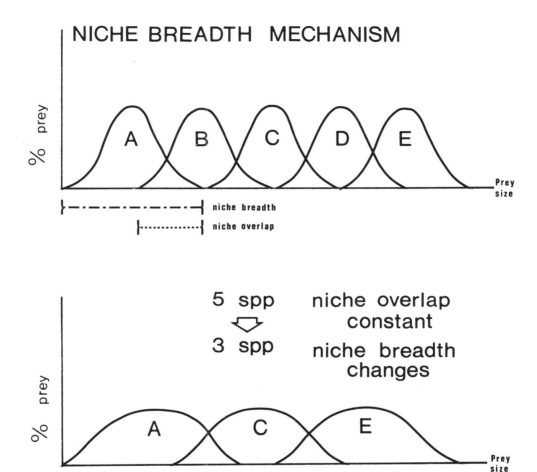

FIG. 18. Schematic representation of the niche-breadth mechanism.

One example of niche-breadth data for kingfishers is given in Table 16. Niche breadth was measured by prey length in the three- and five-species communities. One species, *torquata*, showed an increase in niche breadth in the less diverse community; but the other two, *amazona* and *americana*, showed the opposite, contrary to the predictions of the niche-breadth hypothesis. Mean niche breadths in the two communities were virtually identical, again contrary to prediction of the hypothesis.

TABLE 16. Niche breadths of kingfishers based on prey lengths in the five- and three-species communities.[1]

Species	3-species community	5-species community
torquata	8.33	6.90
amazona	3.85	6.89
inda	-	3.51
americana	2.02	4.76
aenea	-	1.91
mean	4.73	4.79

[1] Frequency distributions of prey lengths were used for niche breadth comparisons, following the formula:
$$\text{niche breadth} = 1/\Sigma \,(P_{ij})^2$$
where P_i is the proportion of diet measured over j prey length categories (Levins 1968). For the calculations, a sample of 20 fishes was selected randomly from each species' diet to equalize differences in sample sizes of prey items.

Severe limitations on comparing niche breadths among species and among communities make interpretation difficult in many cases (Colwell and Futuyma 1971). In the kingfishers, however, the focus is on comparison of niche breadths for the same species between two communities, minimizing these problems. However, because prey-size distributions also changed between the two communities, conclusions concerning changes in niche breadth should be considered tenuous, especially in view of other complications in interpretation of niche breadths depending on the degree to which interference competition plays a role, whether resource availability changes, whether resources are limited, and which measure of niche breadth is used (Williams and Batzli 1979, Feinsinger et al. 1981, Petraitis 1981, Smith 1982). As advocated by Conner and Simberloff (1979) and Thomson and Rusterholz (1982), I have presented my data in such a way that others can recalculate and manipulate niche breadths for reanalyses of the kingfisher data.

There are few quantitative data comparing avian niche breadths along a latitudinal gradient. Askins' (1983) comparison of temperate and tropical woodpeckers also found no consistent differences in niche breadths between temperate and tropical species, although this was consistent with Askins' prediction that no differences would be found under the assumption that woodpecker resources are no less stable in the temperate zone than in the tropics. Terborgh and Weske (1969) and Karr (1971) inferred from mist-net capture data that tropical forest birds show stricter vertical stratification from temperate forest birds. Stiles (1978) found that niche breadths of insectivorous birds in Costa Rica were narrower than those of their counterparts in Washington.

NICHE-OVERLAP MECHANISM

The niche-overlap mechanism is represented schematically in Figure 19, where niche overlap is represented in one dimension by the distance on the horizontal axis in common between resource-use curves of two species, i.e., the range of sizes of the resource used by both species. Niche overlap is a measure of ecological similarity, quantifying to what extent two species use the same resources.

In Figure 19, the number of species is changed from five to three by a decrease in niche overlap without any change niche breadth or resource base. A viewpoint based on interspecific competition would interpret this to mean that a species can tolerate less niche overlap in the less diverse community. Such decreased tolerance of niche overlap could be because predation rates are lower in the less diverse community, which could allow competitor population levels to reach carrying capacity and thus allow increased competitive interactions. Also, a species could tolerate less overlap in the less diverse community because the resource was less abundant or predictable there, forcing it to "monopolize" a greater percentage of the resource base. If competition does not influence community structure, then niche-overlap values should no show consistent among-species or among-communities patterns.

The hypothetical community in Figure 19 competes along only one niche axis. In reality, series of competing species, especially with as many as five species, usually segregate along more than one dimension or niche axis (Schoener 1974). How can overlap values along the various axes be combined to give one measure of niche overlap? Ideally, overlap in multidimensional space can be calculated directly from the data. In other words, if a series of competitors separate along three axes, such as prey size, feeding site, and habitat, and if the habitat and feeding site are known for each prey item, then the volume in three-dimensional niche-space in which two species overlap can be calculated directly (May 1975). Unfortunately, in reality most data sets do not allow such calculations because of low sample sizes in individual data-cells. Then the question becomes: How can the true overlap be estimated from overlap values on the three separate dimensions? This depends on the relationship between the distribution of resources along each dimension. If the distributions are statistically independent, then the percentage of overlap values on each axis can be multiplied to estimate total niche overlap; and if they are highly dependent, then they can be added to estimate total niche overlap (Cody 1974). In practice, the axes will probably be neither completely independent nor dependent, and the multiplicative and additive indices will be low and high estimates, respectively, of true total niche overlap (May 1975).

The niche axes on which most bird species segregate are food, feeding site, and habitat, with habitat referring to some "horizontal" measure of areal overlap, and feeding site referring to specific microhabitat within the habitats. Because data sets are seldom detailed enough to measure overlap on all three axes, other "short-cuts" have been used as estimators, especially for food and feeding site. Cody (1974) used bill dimensions to estimate food overlap, although this has been shown to be unreliable in some cases (e.g.,

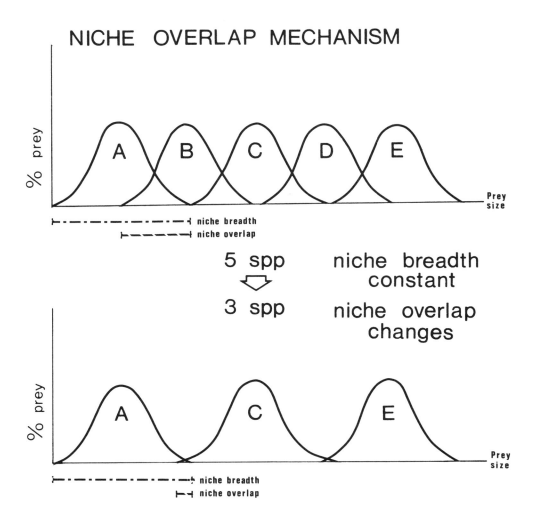

FIG. 19. Schematic representation of the niche-overlap mechanism.

Hespenheide 1975). In kingfishers, bill length was an excellent predictor of prey length; but because prey length itself was estimated, the use of the more indirect bill-length measure was not necessary.

Cody (1974) also used various measurements of the rate of movement to estimate differences in food, but this seems so indirect that its value is questionable. Before such measurements are used, it is necessary to show that: (a) differences in rate of movement result in different prey items taken, and (b) the variance in this dimension for individuals of the same species or for different species is not so great that comparisons using a single estimate of rate of movement are dangerous.

Although all three parameters, food, feeding site, and habitat, were measured for each kingfisher observation, thereby in principle allowing direct calculation of overlap in three dimensions, the sample sizes of prey items were so small that use of specific prey-size distributions for each feeding site in each habitat was not possible. In the absence of larger samples, it seemed more realistic to use a more indirect measure of total overlap, either multiplicative or additive, and to estimate dependence between the axes. There were no statistically significant differences for any species in prey-length frequency distributions among various habitats. There were also no significant correlations between dive entry point (feeding-site selection) and prey size except the following: a significant, positive relationship for *amazona* at Tumi Chucua ($P < .01$, $r = .541$) and Salsipuedes ($P < .01$, $r = .395$), and a significant but negative relationship for *amazona* at ISS ($P < .01$, $r = -.714$). With an increase in the proportion of large fishes with increasing distance from shore, prey length was anticipated to be partly dependent on dive entry point, but this was not the case. There was some dependence between habitat and dive entry from shore in that the opportunities to dive greater distances from shore were fewer in recessed pools and canopy-closed streams. Thus, the verdict on the dependency between the three parameters for kingfishers is ambiguous, and so both multiplicative and additive measures were calculated.

Prey length (Fig. 15) was the parameter used to quantify overlap in food. Differences in taxonomic composition of diets (Table 15) were due primarily to unequal representation of the various taxonomic categories in the prey-size categories. For example, characids tended to become less important in the diet of larger kingfisher species, but this was because the proportion of characids decreased with increasing fish size (Table 4). The reverse trend was evident in Cichlidae. Even differences in frequency of invertebrate prey may be related to prey size: the shrimp taken by *inda* and *americana* were extremely close in length to the mean lengths of the fishes in their diets, and the hard-shelled crabs taken by *torquata* can probably be handled only by that species. Correlations between prey size and prey type are prevalent among studies of insectivorous (Hespenheide 1975) and piscivorous bird diets (e.g., Ashmole 1968, Bédard 1969, Willard 1977).

Distance of dive entry point from shore (Table 14) was used as the feeding-site parameter. Whereas in terrestrial bird communities feeding-site selection is often an extremely important parameter, e.g., foliage, bark, and all their subdivisions are extremely important in ecological segregation (e.g., MacArthur 1958, Root 1967, 1969), in the kingfisher community little separation was evident in feeding-site selection. As previously argued, this was to be expected, because the prey population was highly concentrated in the first 3 ft. from shore, and because the fish themselves are probably much more mobile with respect to distance from shore than are terrestrial arthropods between the various divisions of feeding-site selection. However, when the piscivorous bird community as a whole is considered (see Other Piscivorous Birds chapter above), feeding-site selection becomes more important.

Because the "proportional similarity" measure of overlap (Schoener 1970, Abrams 1980) has been found to be the most reliable measure of overlap over the range of overlap values in the kingfisher data (Linton et al. 1981), it was used to calculate niche overlaps between all species pairs in the community. Overlaps in food were relatively low, overlaps in feeding site were extremely high, and overlaps in habitat use (calculated from Table 11) were low between shaded- and open-habitat species groups, but high within each group. Distributions of resource-use curves for prey length for the five species resembled that for the hypothetical community of Figure 20, but the distributions for feeding site and habitat would be quite different. Overlap values between each kingfisher for each of the three parameters were then multiplied together (Table 17) or added together (Table 18), and then summed to give total niche-overlap (hereafter TNO) values between each possible species pair. Although mean overlap values are also reported in Tables 17 and 18, they are not used in subsequent analyses because of problems with use of mean values as outlined by Thomson and Rusterholz (1982). TNO values are here reported to three decimal places, to correspond to those of Cody (1974) and others. I believe, however, that this degree of precision is misleading and unwarranted, in view of the many methodological uncertainties in this (and every other such) study (see also James and McCulloch 1985). Because the three axes of overlap are not qualitatively distinct resources in the sense of Holt (1987), i.e., all three are direct or indirect measures of overlap in feeding, I believe that I have avoided the problems in interpretation of TNO values outlined by Holt.

The niche-overlap mechanism predicts that niche overlaps will be greater among syntopic species in the more diverse community (Klopfer and MacArthur 1961, MacArthur and Levins 1967, MacArthur 1972, May 1973). This prediction is partially supported by the data. Using either multiplicative or additive indices, the sums of TNO values for each species, i.e., the total amount of niche overlap experienced with all its competitors, are significantly lower in the three-species community (one-tailed Mann-Whitney U-test, $P < .05$). Thus, each species in the five-species community overlaps more heavily with other kingfishers taken together than in the three-species community. The mean TNO values for each species with each of its competitors are lower in the three-species community, using the multiplicative index (.488 vs. .656), but not using the additive index (.708 vs. .558). The decrease in TNO for *torquata* from the five- to three-species community accounts for most of the decrease in multiplicative niche overlap. Note that one cannot look solely at pair-wise overlaps; the overlap with all community members combined must also be considered.

When TNO values between adjacent species in the size continuum are examined, no consistent pattern is evident, with both high and low values found. Consideration only of values between adjacent species gives a chaotic picture of community structure. However, when the sums of the TNO values for each species with all its potential competitors are analyzed, a remarkable consistency is evident. In the five-species community, four of the five species' TNO values are very close (.665 to .757) with one outlier (.489) using the multiplicative index (Table 17), and using the additive index (Table 18), the TNO values range from 2.016 to 2.519. In the three species community, the TNO values using the additive index are very similar (1.364, 1.423, 1.461), but no such similarity is evident using the multiplicative index.

TABLE 17. Overall niche-overlap values using multiplicative index.

Five-species community

Species	torquata	amazona	inda	americana	aenea
torquata	—	.530	.028	.097	.010
amazona	.530	—	.045	.115	.011
inda	.028	.045	—	.335	.258
americana	.097	.115	.335	—	.210
aenea	.010	.011	.258	.210	—
Sum	.665	.701	.666	.757	.489
Mean	.166	.175	.167	.189	.122

Three-species community

Species	torquata	amazona	americana
torquata	—	.244	.104
amazona	.244	—	.384
americana	.104	.384	—
Sum	.348	.628	.488
Mean	.174	.314	.244

To examine whether the sums of TNO values were more constant than expected by chance, I randomly redistributed the 10 TNO values from the five-species community in Table 17 among the 10 species-overlap pairs 25 times, and then calculated the variances of the 25 resulting sets of five TNO sums. The mean variance of the artificial communities was 0.0653 (range 0.0127 to 0.1260), and only three of the 25 randomly produced variances were lower than the observed variance (0.0270). Thus, the probability that the low variance observed among the TNO values in the five species community was produced by chance is about .12-.16. The same exercise repeated for the TNO values using the additive index (Table 18) found that the variance of five observed values was lower than all but 5 of 25 artificial communities; thus, the probability that the observed TNO values using the additive index were less heterogeneous than expected by chance is about .20-.24.

TABLE 18. Overall niche-overlap values using additive index.

Five-species community

Species	torquata	amazona	inda	americana	aenea
torquata	—	.829	.407	.579	.350
amazona	.829	—	.438	.594	.348
inda	.407	.438	—	.721	.693
americana	.579	.594	.721	—	.625
aenea	.350	.348	.693	.625	—
Sum	2.165	2.209	2.259	2.519	2.016
Mean	.541	.552	.565	.630	.504

Three-species community

Species	torquata	amazona	americana
torquata	—	.700	.664
amazona	.700	—	.761
americana	.664	.761	—
Sum	1.364	1.461	1.423
Mean	.682	.731	.712

The use of the observed TNO values, however, is a highly conservative way to generate hypothetical communities that biases the outcome in favor of acceptance of the null hypothesis (Grant and Abbott 1980). TNO values theoretically can range from 0 to 1, rather than within the more limited range of empirically determined values. If 25 new communities of 5 species are created, with pair-wise overlaps randomly chosen between 0 and 1.00, the TNO values then calculated from these 25 "communities" are much more variable than those observed in the five-species kingfisher community. Using the multiplicative index, the mean TNO value of the 25 random groups is .502 (s.d. = .287, range .151-1.23) and the mean coefficient of variation of the 25 sets of five TNO values is

.491 (s.d. = .128, range .278-.821). The coefficient of variation of the five real-world TNO values is .153, much lower than any of the random communities. Therefore, the real TNO values are less variable than expected by chance (< 1/25, $P < .04$). Using the additive index, the results are similar, but the difference between real and random communities is less dramatic. The mean TNO value of the 25 random groups is 5.71 (s.d. = .808, range 4.38-7.56; none is as low as the mean of real-world group, 2.23) and the mean coefficient of variation is .082 (s.d = .043, range .051-.210). Three of the random groups have lower c.v.'s than the real-world group, and so the probability that the real world group's observed TNO values are less heterogeneous than expected by chance is about .12-.16.

My interpretation of the real vs. random comparisons is that the low variation among real TNO values in the five-species community is consistent with predictions of the hypotheses of "limiting similarity" (MacArthur and Levins 1967) and "diffuse competition" (MacArthur 1972). Between-species TNO values seem to be complementary within a community. For instance, the TNO between one set of adjacent species, *torquata* and *amazona*, is high using either overlap index, but the TNO values for these two species with the other three species are low, resulting in TNO sums close to the mean. Similarly, TNO between adjacent species *amazona* and *inda* is low, but this is compensated for by high TNO values with other community members, again resulting in a TNO sum close to the mean. However, with only one of the four comparisons of the observed pattern with randomly created patterns statistically significant in terms of standard P values ($P < .05$), the support for limiting similarity is weak. Use of such indirect means of searching for the influence of interspecific competition on "structuring" communities is less than satisfying, especially with a sample size of only one real-world community. Also, the evidence for limiting similarity breaks down in the results from the three-species community, in which, using the multiplicative index, this limiting similarity plunges for *torquata* but not for *amazona* or *americana*.

Comparison of the five- and three-species communities (Table 17) shows that the smallest species in both the five- and three-species communities have virtually identical TNO values. The intermediate species in the three-species system is reasonably close to the other TNO values in the five-species system. Thus, it is the nearly 50% decrease in TNO for *torquata* that is responsible for nearly all the difference between the five- and three-species communities.

Does the niche-overlap mechanism really explain the proximate mechanism of change in species number between the five- and three-species communities? Because the difference between sums of TNO values in the five- vs. three-species communities using the multiplicative indices would not be significant if the TNO sum for *americana* were a mere 0.002 higher, or if *torquata* were not included in the analysis, it could be argued that the niche-overlap mechanism is only marginally supported by the data. An alternative interpretation, however, that supports the niche-overlap mechanism is that the apparent decrease in tolerance of TNO by *torquata* in the three-species system, in which *amazona* then feeds on smaller fish, is the catalyst for the loss of two species.

Pianka (1972) predicted that mean non-zero niche overlap within communities should be negatively correlated with species richness, assuming that intensity of competition is positively related to number of species. Niche-overlap data from the two kingfisher communities (Tables 17 and 18) support this hypothesis: using either multiplicative or additive indices, mean niche overlap is lower in the five-species than in the three-species community. Similar results were obtained for lizards by Pianka (1974), for desert rodents by M'Closkey (1978), and for insectivorous birds in pine forest by Rusterholz (1981a).

Using either overlap index in the five-species community, the smallest species, *aenea*, shows the lowest TNO values. A tempting, post-hoc explanation for low TNO values for the smallest species is that perhaps species at the bottom end of the size spectrum "need" to maintain a lower TNO because of their potentially precarious position: fishes smaller than those that they are already taking are extremely rare, and if they are under competitive pressure from other species during environmental fluctuations, their persistence in the community may be jeopardized. When the prediction of low TNO values for the smallest species is applied to the three-species community, however, the prediction is not upheld: *americana* does not have the lowest TNO values.

The species with the highest TNO sum in the five-species system, *americana*, is also the first species to drop out of the system during times of low resource availability, i.e., high-water season. This is similar to results obtained from a group of insectivorous birds by Rusterholz (1981b). Thus, perhaps these TNO values can be used as an index of competition for resource increases. On the other hand, although *americana* is the species that disappears seasonally, *inda* and *aenea* are the first two species to disappear permanently on a latitudinal or elevational gradient, and this certainly could not be predicted from their TNO values in the five-species system, particularly for *aenea*. In the three-species community, the species with the highest TNO value, *amazona*, is the next species to drop out latitudinally; but with only three species in the community, little significance can be attached to this result.

Because most of the above discussion depends on a highly derived measure (TNO is two steps removed from original prey data), not much faith can be placed in the conclusions from a single study such as this one, especially with the number of between-site, uncontrollable variables (e.g., prey size-distribution and abundance, habitat structure, predator species). Such analyses need to be repeated for a variety of systems to see if similar patterns emerge; only then can single-study results properly be appraised. Furthermore, the numerous statistical and interpretational problems associated with measures of niche overlap, controversy over which measure of overlap is best, and even whether niche overlap is at all related to interspecific competition (e.g., Hanski 1978; Hurlbert 1978, 1982; Petraitis 1979; Abrams 1980; Ricklefs and Lau 1980; Slobodchikoff and Schulz 1980; Linton et al. 1981; Maurer 1982; Smith and Zaret 1982; Thomson and Rusterholz 1982; Diamond 1984; Smith 1984; Glasser and Price 1988) undermine confidence in such analyses. At this point, I wonder whether any of the metrics presented in Tables 16-18 have any real meaning. I join the ranks of ecologists who are frustrated by our inadequate understanding of the underpinnings of the quantitative study of interspecific competition and its associated metrics (Glasser and Price 1988). In any case, as with niche breadths, I have presented my data in such a way that overlap values can be recalculated using other measures of overlap to examine effects on my results and interpretations.

RESOURCE-BASE MECHANISM

Resource base is here represented schematically in one dimension (Fig. 20) by the length of the horizontal axis, i.e., the total range of a resource used by five species. Increasing the resource base by adding another dimension to allow additional separation is another form of this mechanism. In both cases, it is the range of the resource that is increased. Note that changes in the quantity or predictability of a resource within that range used by a species are fundamental to the biological interpretations of the niche-breadth and niche-overlap mechanisms; and in this sense, all three mechanisms are different varieties of the "resource-base mechanism." In the strictest, proximate sense, used in Figure 20, the resource-base mechanism is the addition or subtraction of resource availability with respect to total range used.

In Figure 20, the number of species changes from five to three without any change in niche metrics except a reduction in range of the resource: its availability is reduced such that the two species using the upper end of the resource distribution are forced out of the community. These two species can either disappear, as do "D" and "E" in Figure 20, or, by directing their resource use toward the smaller end of the size distribution, they can replace species that formerly used this portion of the resource, say species "B" and "C."

The change in the resource base (the surface fish population, for kingfishers) between the five- and three-species communities, quantified in Tables 2 and 3, is dramatic in both density and size distribution. Not only does density drop by an order of magnitude, but fish size distribution shifts significantly toward smaller fishes. But does this affect kingfisher prey-size selection? As previously shown (Table 15), two of the three species in the three-species system showed dramatic and statistically significant shifts toward smaller fishes, paralleling the shift in surface fish distribution. My interpretation is that with the increasing rarity of all fishes, especially larger ones, kingfishers can no longer "afford" to wait for larger fishes as they did in the five-species community, and so they shift to small fishes. If the kingfisher community is taken as a whole, with equal weight given to each species, to form a composite diet with respect to fish size-classes, the frequency distribution corresponds closely with that of surface fish sizes; larger fishes are slightly overrepresented in the diet, and smaller fishes are under-represented (Remsen 1978).

An even closer correspondence can be obtained by using the number of schools rather than the number of individual fishes in each size-class. The majority of fishes in any given school are usually in the same inch-size-class, and so each school could be thus categorized. Although use of number of schools can be justified on theoretical grounds in looking for correlations between fish density and bird density (see Seasonal Differences, in Kingfisher Density chapter), the correlation is not improved by using number of schools rather than individuals. However, plotting composite kingfisher diet with number of schools for the spectrum of fish size-categories (Fig. 21) produces two plots remarkably

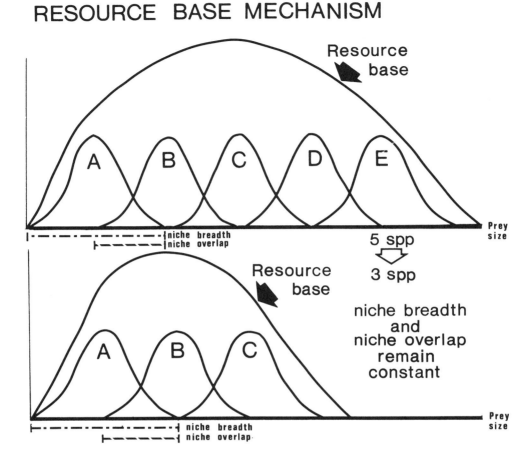

FIG. 20. Schematic representation of the resource-base mechanism.

similar and statistically indistinguishable (Kolmogorov-Smirnov test, $P > .05$). This is true for both five- and three-species communities. The way that kingfisher diet tracks fish size distribution is impressive and, short of experimental manipulation of fish populations, is my best evidence for the importance of the resource base in structuring kingfisher communities.

How does the change in resource base specifically affect kingfishers? The one-dimensional resource-base mechanism is not operating, because no change in the range of prey size is involved. (The largest and smallest fishes taken by any kingfisher in the five-

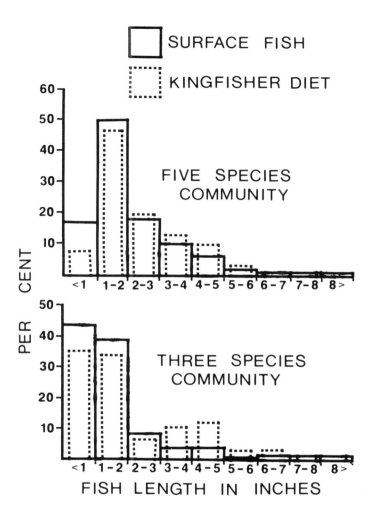

FIG. 21. Kingfisher diet vs. surface fish densities for various fish size-classes. Each kingfisher species was weighted equally in the calculations.

species community are also being taken by kingfishers in the three-species system.) The resource-base mechanism, however, is operating when the second dimension of ecological separation, habitat preference, is considered: the two species that dropped out of the system separated to a significant degree from the remaining three on the basis of habitat (Table 11).

This dimension, however, is not used at all in the three-species community. Habitat suitable from a structural basis is certainly present at Salsipuedes. Why is it not used, even by the three species present? Unfortunately, surface fish density in shaded habitat at Salsipuedes was not quantified sufficiently to permit statistical comparisons. Only six surface fish censuses were taken there, and not a single fish was recorded. Surface fish density there was obviously extremely low. For comparison, the probability of obtaining six consecutive negative counts in open habitat at Salsipuedes was approximately .0000002. Thus, it is safe to say that surface fish density was definitely lower in shaded habitat, perhaps to the same degree that open habitat had higher densities than shaded habitat at Tucuchira. The only plausible explanation for the lack of *inda* or *aenea* in shaded habitats at Salsipuedes seems to be that surface fish were so scarce that foraging there was not energetically feasible. This hypothesis is supported by the lack of expansion into shaded habitats by the other three species.

Thus, my interpretation of these results is that *inda* and *aenea* seem to be forced out of the system by a combination of the resource-base and niche-overlap mechanisms. Fish density in their preferred habitat drops so low that it is unable to support any kingfishers. They cannot shift their habitat preferences to open habitats, because *americana* and *amazona* occupy the open habitats and have shifted their prey-size preferences downward and are, therefore, taking fish the mean size of which is smaller than those taken by *inda* and *aenea* in the five-species system. Thus, *americana* would be overlapping with *aenea*, and *amazona* with *inda*, to a much higher degree than can be tolerated. Part of the impetus for the downward shift may come from decreased tolerance of niche overlap by *torquata*, which, although taking fish of the same size as it took in the five-species system, evidently can no longer tolerate as much overlap, because larger fish are now disproportionately rare. The decrease in the density and mean size of the surface fish population is the single most important factor in determining the number of species that can coexist.

An alternative explanation is that *inda* and *aenea* drop out of the system in the savannas for some reason not related to the change in fish density and size distribution, and that their absence then allows *amazona* and *americana* to undergo the downward shift in prey-size selection. An assumption of this hypothesis is that taking smaller fish is more advantageous. The arguments against this have been outlined earlier in the Prey Size section. Stated succinctly, this could be the case only if small fishes were much easier to catch than larger fishes, and there was no evidence to indicate this was (or was not) so. I also cannot think of any other reason why *aenea* or *inda* would drop out of the system. For instance, their preferred habitat is extensive.

But why do *inda* and *aenea* rather than *amazona* and *americana* drop out of the system? After all, *amazona* and *americana* are in a sense "intruding" into the fish size-class that are the specialties of *inda* and *aenea*, respectively, and these latter two should theoretically be more efficient at capturing these fishes and have the competitive advantage. But *amazona* and *americana* also have some advantages that evidently "tip the scales" in their favor. Because shaded habitat seems to become unsuitable for kingfishers, *inda* and *aenea* would have to shift their foraging to more open habitats, where *amazona* and *americana* may be more efficient. Considering the importance of sunniness and shadiness in dividing kingfisher habitats, perhaps skills required for fishing in normally sunlit water differ from those for fishing in normally shaded water. Second, although *amazona* and *americana* may not be as efficient at catching smaller fishes, they can eat fishes larger than *inda* and *aenea* can take, respectively. *Amazona* and *americana* can use large fish sizes that their smaller counterparts cannot eat, whereas they can eat all the fish sizes eaten by *inda* and *aenea*. Theoretical considerations of this advantage have been outlined by Wilson (1975). Third, if any of the competitive exclusion is mediated by interference competition, the larger species would always be the winner (Table 5).

Following this reasoning, however, why doesn't *inda* then exclude *americana* rather than allowing itself to be ousted by *amazona*? It is larger than *americana*, giving it the advantages for the larger of a species pair. First, the differences in adaptation to foraging in open vs. shaded habitat may be more important. Second, if size were always the most important factor, species should drop out latitudinally according to size, the smallest first, and this is clearly not the case. If *inda* were to replace *americana* in the three-species system, it would be taking fishes "two notches" below those it was taking in the five-species system, catching fish smaller than those caught by *aenea*, which is two sizes smaller than *inda*. Perhaps differences in foraging efficiency are not so great when a species shifts downward one size-class into that used by the next-smallest species, but the reduction in efficiency in shifting more than one size-class downward may be so great that the advantages of larger size are not enough to offset the loss in efficiency. Experimental manipulation of prey populations in laboratory populations of kingfishers would provide critical data for testing these hypotheses.

Why do surface fish populations showed reduced densities and smaller sizes in the same savannas? This was obviously beyond the scope of the present study, and only speculation is possible. Certainly the greater amplitude of water-level fluctuations on a daily, seasonal, and possibly annual basis in the savannas must have a depressing effect on fish densities and diversities (see Study Sites chapter). If mean life expectancy is reduced because of this greater unpredictability, then mean fish size would also decrease, all else being equal. Smaller fish may also have an advantage in the low-water season, when amount and depth of remaining water is reduced (Kushlan 1974, 1976a). A combination of climatic factors is probably responsible for the differences in surface fish population characteristics, and these climatic factors would then be the ultimate determinants of kingfisher diversity.

DISCUSSION

ASSUMPTIONS AND CONCLUSIONS

An underlying assumption in the portion of this study that deals with community structure has been that competition among kingfishers strongly influenced most parameters measured. I have presented only weak evidence for interspecific competition per se. Caswell (1976), Wiens (1977), Strong et al. (1979), and many others have questioned whether interspecific competition plays any part in shaping community properties or differences observed between species. However, tabulations of experimental results shows that evidence for interspecific competition was found in 90% of over 150 studies (Schoener 1983) or 40% of 120 studies (Connell 1983).

Although a review of the current controversy over the prevalence of effects of competition on community structure is obviously beyond the scope of this study, a few points will be offered here. A major objection to the generality of competition concerns the absence of a demonstrably limited resource. However, if it can be granted that time itself is always limited, and that how time is budgeted is a parameter sensitive to natural selection, then this objection can be partially satisfied. Thus, any interspecific interactions that adversely affect time-energy budgets could be called competitive interactions. For example, when food itself is unlimited, if one species removes food in such a way that it takes another species longer to obtain the same amount of food than it would have had the first been absent, then this represents interspecific competition, because the increased time required could have been budgeted toward other fitness-increasing activities. Thus, virtually any ecological overlap could fit this definition of potential competition. The problem, of course, is to demonstrate that such effects exist.

As for the importance of competition in shaping kingfisher community structure, the verdict awaits the outcome of experimental manipulations that could not be performed in the current study.

A number of major weaknesses in this study, several of which have been discussed previously, make generalizations from the results somewhat tenuous. The niche parame-

Discussion

ters should have been measured for at least one full annual cycle, especially in the savannas, where seasonal differences would be predicted to be more extreme. Assuming that the results from November and December are applicable to the entire system is dangerous, although this period displays nearly the full range in water-level fluctuations. Investigation of between-year differences was beyond the scope of this study.

Another weakness is that for three-species systems, and comparison of these with five-species systems, the conclusions of this study rest on data from only one site. Although the close similarities between data from the five-species sites were encouraging, greater between-site variability might be expected between different three-species sites. I cannot rule out the possibility that the reduced densities and fish lengths at Salsipuedes were peculiar to that site; data from savanna fish faunas elsewhere, however, parallel my findings with respect to both density and body size (McConnell 1975). Also, as mentioned earlier, data are needed from sites with other combinations of kingfisher species. Given the intensive efforts needed to gather the data for this study, an investigator wishing to pursue this problem will need either to devote many years to gathering the field data or to be able to muster a team of investigators.

If the results of this study can be generalized to the kingfisher system as a whole, the following predictions can be made: (1) surface fish density determines the number of coexisting kingfisher species; (2) surface fish density declines along a latitudinal gradient; (3) as fish density decreases, tolerance of total niche overlap will decrease, forcing species out of the system as species richness declines from three to two to one; and (4) the number of species at any one site will reflect the most efficient way to use resources: thus the two-species system, with *torquata* and *americana*, gives way to the single-species system, *alcyon* alone, at that point where surface fish density and size-distribution shifts are such that an intermediate species outcompetes one larger and one smaller species (MacArthur 1972). Terborgh (1980) pointed out that the temperate-zone member of a guild was often near the midpoint of the body sizes of the tropical members of the comparable guild. Perhaps anomalous situations such as the absence of *amazona* from areas where *inda* or *aenea* or both are present (e.g., western Ecuador and Yucatan) will be explained by anomalous fish size-distribution and density patterns. Documentation of surface fish densities and kingfisher diets in these various species combinations would be relatively simple. Convincing experimental manipulations of the fish resource-base, however, would be difficult to implement, and the short-term results difficult to interpret.

An assumption implicit in my discussion of the geographic ecology of kingfishers, up to this point, has been that their distributions are not limited in any significant way by barriers to dispersal. This assumption is probably valid, especially for the three open-habitat species, *torquata*, *amazona*, and *americana*, all of which have shown the ability to cross barriers of unsuitable habitat via true dispersal and not "vicariance." Also, *inda* and *aenea* drop out latitudinally long before any substantial barriers to dispersal are encountered.

If the results of this study are applicable to avian communities as a whole, the resource-base mechanism (increasing dimensionality) and the niche-overlap mechanism, in response to declining resource-base, will be the primary proximate factors involved in species-packing in latitudinal gradients of species diversity. The addition of resources not available in the temperate zone, such as year-round supply of fruits, nectar, and large insects, as well as the presence of certain habitats and microhabitats not found or in the temperate zone, accounts for a substantial portion of the increased species richness of the tropics (Orians 1969; Schoener 1971; Karr 1971, 1975; Stiles 1978; Terborgh 1980; Askins 1983; Remsen and Parker 1983, 1984; Remsen 1985), and differences in resource availability have been postulated as partly responsible for differences in diversity between tropical forests in Africa and South America (Karr 1976, 1980; Pearson 1977). As for fish-eating birds, Kushlan et al. (1985), in comparing two areas with different numbers of species of piscivorous birds, found that differences in resource bases were an important determinant of species richness: the more tropical site had a larger standing crop of fishes, a greater percentage of large fishes, and a larger number of habitat types used by fish-eating birds.

Why can't a sixth species be added to the five-species kingfisher community? If there is a limit to total niche overlap, as argued earlier, then insertion of a sixth species into the existing five-species community would result in niche-overlap values above this threshold, unless the existing species' niche parameters were altered to accommodate the newcomer. Hypothetically, the existing species could shift prey-size preferences upward or downward — but this would force the smallest species, *aenea*, to feed on a resource virtually nonexistent on the water surface, i.e., fishes smaller than 12 mm; or it would force the largest species, *torquata*, not only to shift to increasingly rarer size-classes of larger fishes, but probably to increase overlap with larger piscivorous birds, such as herons and *Pandion haliaetus*, that take these larger fishes. These same arguments are applicable to why a sixth species could not be added on to either end of the kingfisher size-sequence, i.e., a species smaller than *aenea* or one larger than *torquata*.

If it is impossible to add a sixth species by expansion along one niche axis, i.e., food, then could a sixth species be accommodated by expansion along other niche axes, i.e., microhabitat or habitat selection? I see no "unoccupied niches" available to piscivorous kingfishers. For example, foraging at greater distances from shore and employing extensive hovering, as in the Pied Kingfisher *Ceryle rudis* (Junor 1972, Douthwaite 1976, Draulans et al. 1981, Johnston 1989), is a possibility, yet the presence of two terns and *Pandion haliaetus* in this offshore zone may preclude further niche expansion in that direction. Ecological equivalents of these species seem to absent on the lakes where *Ceryle rudis* fishes extensively offshore.

HYPOTHESES CONCERNING TROPICAL SPECIES DIVERSITY

Although this study was concerned primarily with the proximate factors involved in tropical species-packing, the kingfisher data can be used to evaluate the six hypotheses reviewed by Pianka (1966) that have been proposed as "subultimate" causes. (See also McConnell [1975] for review of these hypotheses with respect to latitudinal gradients in fish diversity).

(1) *Diversification with Time*: This hypothesis proposes that all communities diversify with time, never reaching an equilibrium point; the temperate zone is considered to be the "younger" community because climatic disturbances are more severe and more frequent, and because temperate climates are relatively recent in origin. Intercontinental comparisons

of kingfisher diversity provide evidence against the theory that communities diversify with time.

The center of kingfisher diversity is the Old World, particularly the Australasian region (Fry 1980a, Forshaw 1983). New Guinea, with 18 species in 6 genera, is clearly the richest area on earth in kingfisher density (and these totals exclude species found on nearby small islands). New Guinea has two endemic genera (*Clytoceyx*, *Melidora*), and two more genera are restricted to the New Guinea-Australia area (*Tanysiptera*, *Dacelo*). Southeast Asia and Indonesia (16 species in 7 genera, including one endemic genus, *Lacedo*) and the Philippines (13 species in 4 genera) are secondary centers of diversification. Africa is also rich in kingfishers, with 16 species in six genera, including one endemic genus, *Corythornis*. In contrast, the entire New World has only six species in only two weakly differentiated genera, one of which (*Ceryle*) is also found in the Old World. Clearly, by conventional zoogeographic and taxonomic standards, the center of kingfisher evolution has not been in the New World but somewhere in the Old World, possibly New Guinea. Kingfishers have in all likelihood been present in the Old World millions of years longer than in the New World. Although the published fossil record is still scanty, Old World kingfisher fossils date back to at least the Eo-Oligocene (Mourer-Chauviré 1982), but New World kingfisher fossil specimens are no older than the late Pleistocene and are all clearly referable to neospecies (Brodkorb 1971).

Despite their relatively recent arrival in the New World, involving no more than two invasions (Fry 1980a), kingfishers have diversified there to span completely the entire range of sizes displayed by Old World kingfishers, at least among the aquatic forms. The Old World's largest fish-eating kingfishers, *Ceryle maxima* and *Pelargopsis capensis*, are just slightly larger than *C. torquata* of the New World, and the smallest Old World piscivorous kingfishers, such as *Corythornis leucogaster* and *Alcyone pusilla*, are the same size as *Chloroceryle aenea* of the New World. Furthermore, when only piscivorous species are included, the maximum species-packing situation in the New World, 5 species, is the same as that for Africa (Table 19) and higher than anywhere in Asia or Australasia (Bell 1981, Forshaw 1983, Simberloff and Remsen, in prep.).

TABLE 19. Bill lengths of piscivorous kingfishers of the Amazon and Congo basins. Bill lengths (exposed culmen) for Congo species taken from 10-21 specimens of each species.

Amazon	Mean bill length (mm)		Congo
Ceryle torquata	75	77	*Ceryle maxima*
Chloroceryle amazona	64	56	*Ceryle rudis*
Chloroceryle inda	47	42	*Alcedo quadribrachys*
Chloroceryle americana	39	31	*Corythornis cristata*
Chloroceryle aenea	28	28	*Corythornis leucogaster*

Therefore, Old World fish-eating kingfishers do not display greater species-packing than their New World counterparts, contrary to the predictions of the Time hypothesis. The remarkable similarities in the size series in the Amazon and Congo basins suggest not only that the "kingfisher niche" is already saturated in the New World, but that kingfisher communities may show convergent community evolution (Remsen, in prep.). Furthermore, descriptions of the preferred habitats of the middle-sized species, *Alcedo quadribrachys*, in the five-species community in western Africa are remarkably similar to those of its size-counterpart in the New World, *Chloroceryle inda*; *A. quadribrachys* "perches over water shielded by foliage" and "along brooks only a few yards wide" (Chapin 1939:292), and is "easily overlooked as it sits motionless on a low, secluded branch overhanging water" (Forshaw 1983). *Alcedo quadribrachys*, with its extensively rufous underparts, also looks like a blue-backed male *C. inda*. The smallest species in both western Africa and Amazonia, *Corythornis leucogaster* and *Chloroceryle inda*, also resemble each other in the mixture of rufous and white coloration on the underparts. Such similarities suggest convergence beyond similarity in bill length and relative position in the kingfisher community. The comparison with respect to bill ratios, and controversy as to their relevance to community structure (Strong et al. 1979, Grant and Abbott 1980, Wiens 1982), will be discussed elsewhere (Remsen, in prep.).

If tropical species diversity was nothing more than the product of a greater excess of speciation events over extinctions, then the three species-packing mechanisms, all of which depend on a causal relationship between resource use and diversity, are irrelevant. Therefore, such a hypothesis predicts that no particular pattern would be expected from niche metric data. Although the niche breadth data for kingfishers showed no particular pattern, both the niche-overlap and resource-base calculations showed non-random patterns of change between communities.

(2) *Spatial Heterogeneity*: This hypothesis explains the increase in species diversity in the tropics by invoking an increase in spatial heterogeneity with decreasing latitude. Kingfishers were chosen for study specifically to minimize this problem, as discussed previously. The two species most restricted to the tropics, *aenea* and *inda*, are the two that show habitat separation from the other three (Table 11); yet, if the situation at Salsipuedes is typical, then it is the decline of a resource within their shaded habitat, not disappearance of the habitat itself, that is the important factor. The habitat itself is still available. Habitat suitable for both species extends farther north in Middle America than does the distribution of either species (E. O. Willis, in litt.). Completely shaded habitat structurally similar to that favored by *inda* and *aenea* is also available extensively in the Gulf Coast lowlands of the southeastern United States (personal observation). Thus, the Spatial Heterogeneity hypothesis is not supported by the kingfisher data.

(3) *Competition*: This hypothesis proposes that interspecific competition is more intense in the tropics because of the increased importance there of biological rather than physical factors. As a result, niche breadths are predicted to be smaller in the tropics. My data on niche breadths, presented previously (Table 16), do not support this hypothesis, although the latitudinal range encompassed by the study sites was not sufficiently broad to provide a strong test.

(4) *Predation*: This hypothesis proposes that predation rates are higher in the tropics, depressing population below carrying capacity, thereby lowering the intensity of competition between species and thus permitting the coexistence of more species. In terms of niche metrics, this hypothesis predicts greater niche overlap in the tropics. The kingfisher data fit this prediction in a limited way (see Niche-Overlap Mechanism in preceding chapter), but there was no evidence for increased predation rates in the less diverse community.

The data for increased predation rates on birds in the tropics have come primarily from nest studies. There are insufficient data at present from tropical kingfisher nests to determine whether they fit this pattern. In any case, the generality of increased nest predation rates in the tropics has not been adequately demonstrated. Most data come from areas that have been disturbed to varying degrees and from which larger carnivores have been removed or greatly reduced, perhaps favoring the increase in smaller carnivores more likely to be nest predators (Oniki 1979, Willis and Eisenmann 1979). Preliminary data from less disturbed regions indicate that the dogma concerning higher predation rates on nests of tropical birds merits re-examination (Oniki 1979, Loiselle and Hoppes 1983; Skutch 1985).

Predation rates on adult kingfishers are unknown. These data are virtually impossible to gather, because predation events are so rarely witnessed. Although longevity of tropical birds is in general greater than temperate zone counterparts (Snow and Lill 1974, Loftin 1975, Fry 1980b, Bell 1982), data are not available on relative predation rates in the two regions. Furthermore, greater longevity does not necessarily mean lower predation rates. Predation rates on adult tropical birds could be relatively much greater than on adult temperate birds, yet total mortality rate could be much lower because of the overwhelming importance in the temperate zone of mortality caused by physical factors such as weather fluctuations and hazards of migration. For example, annual adult mortality for a temperate-zone passerine might be 50%, but only 5% of that might from predation; a comparable tropical passerine might have an annual adult mortality rate of 10%, of which 95% might be from predation.

The qualitative impression of density and diversity of potential predators on adult birds in the tropics is that these are greater than in the temperate zone. For example, an Amazonian forest might have five species of felids, several species of opossums and mustelids, a vampire bat feeding on birds, as many as seven species of bird-eating or partly bird-eating raptors (Terborgh et al. 1984), and several species of arboreal, bird-eating snakes. A comparable inventory in the temperate zone would be much, much lower. Species richness of predators, however, cannot be equated with predation rates. Unfortunately, data for predation rates on forest birds may never be obtained. More generally, the few data available so far from the tropics do not support the Predation hypothesis in that tropical bird populations are not below carrying capacity (Montgomerie and Gass 1981, Greenberg and Gradwohl 1986).

(5) *Climatic Stability*: This hypothesis proposes that the greater stability of the climate has allowed greater specialization in the tropics, permitting more species to use the same resource space. It predicts that niche breadths will be smaller in the tropics. Data from this study, however, do not support this hypothesis, as discussed previously, where the same prediction was given in terms of niche metrics (although the Climatic Stability hypothesis

makes different predictions in terms of competition and density of individuals — see Pianka 1966). Another prediction of this hypothesis is greater tolerance of niche overlap in the tropics, because of increased reliability of resources within a niche and consequent diminished need to maintain exclusive areas of niche space. This hypothesis can also be used to predict a greater range of resource availability. Thus, the kingfisher data cannot really be used as evidence for or against this hypothesis. However, climatic stability may affect the kingfisher data indirectly, in that the greater stability of water levels and other climatic factors must play a major role in the difference in prey resource bases at the two types of study sites.

(6) *Productivity*: This hypothesis proposes that increased productivity and absolute abundance of resources as a result of increased solar radiation received at tropical latitudes, coupled with increased rainfall, is responsible for the greater diversity of the tropics. In practice, disentangling this hypothesis from the previous one is difficult (Pianka 1966); thus, it is perhaps best to consider these two together as one hypothesis, because it is probably the combination of the two that is important to latitudinal gradients in species diversity. Certainly the Productivity hypothesis is operating indirectly on the kingfisher data via effects on the resource base: the decreased density of surface fishes in the savannas must be at least partly related to the decreased productivity of the waters there. Furthermore, preliminary data reveal that my tropical sites may be much more productive in terms of resources for kingfishers than temperate areas: the total density of kingfishers at my tropical stream sites was much higher than that at two sites in North America (see Kingfisher Density section).

The predictions of the Productivity hypothesis in terms of niche metrics can involve all three mechanisms: niche breadth can increase, because of increasing abundance of resources within a given category; niche overlap can increase, because the need to maintain exclusive areas of niche space is diminished; and increased productivity can increase the range of resources available. Thus, the niche-metric data from the kingfisher system cannot really be used as evidence for or against this hypothesis.

In summary, the kingfisher data do not support the Diversification with Time hypothesis, the Spatial Heterogeneity hypothesis, or the Competition hypothesis, but are ambiguous concerning the Predation hypothesis, the Climatic Stability hypothesis, and the Productivity hypothesis. The last two are probably the most important determinants of the resource base, surface fishes, and, therefore, may be the ultimate factors governing kingfisher diversity.

Literature Cited

ABRAMS, P.
1980.   Some comments on measuring niche overlap. Ecology 61:44-49.

AINLEY, D. G., D. W. ANDERSON, and P. R. KELLEY.
1981.   Feeding ecology of marine cormorants in southwestern North America. Condor 83:120-131.

ALATALO, R. V., and J. MORENO.
1987.   Body size, interspecific interactions, and use of foraging sites in tits (Paridae). Ecology 68:1773-1777.

AMERICAN ORNITHOLOGISTS' UNION.
1983.   Check-list of North American birds, 6th edition. American Ornithologists' Union, Lawrence, Kansas.

ASHMOLE, N. P.
1968.   Body size, prey size, and ecological segregation in five sympatric tropical terns (Aves: Laridae). Syst. Zool. 17:292-304.

ASKINS, R. A.
1983.   Foraging ecology of temperate-zone and tropical woodpeckers. Ecology 64:945-956.

BALTZ, D. M., G. V. MOREJOHN, and B. S. ANTRIM.
1979.   Size selective predation and food habits of two California terns. West. Birds 10:17-24.

BAYER, R. D.
1985.	Bill length of herons and egrets as an estimator of prey size. Colonial Waterbirds 8:104-109.

BÉDARD, J.
1969.	Adaptive radiation in Alcidae. Ibis 111:189-198.

BELL, H. L.
1981.	Information on New Guinean Kingfishers, Alcedinidae. Ibis 123: 51-61.

1982.	Survival among birds of the understory in lowland rainforest in Papua New Guinea. Corella 6:76-82.

BETTS, B. J., and D. L. BETTS.
1977.	The relation of hunting site changes to hunting success in Green Herons and Green Kingfishers. Condor 79:269-271.

BLAKE, E. R.
1953.	Birds of Mexico. Univ. of Chicago Press, Chicago.

1977.	Manual of Neotropical birds, vol. 1. Univ. of Chicago Press, Chicago.

BOHALL, P. G. and M. W. COLLOPY.
1984.	Seasonal abundance, habitat use, and perch sites of four raptor species in north-central Florida. J. Field Ornith. 55:181-189.

BRITTON, R. H. and M. E. MOSER.
1982.	Size specific predation by herons and its effect on the sex-ratio of natural populations of the mosquito fish *Gambusia affinis* Baird and Girard. Oecologia 53:146-151.

BRODKORB, P.
1971.	Catalogue of fossil birds, part 4 (Columbiformes through Piciformes). Bull. Florida State Mus. 15:163-266.

BROOKS, R. P. and W. J. DAVIS.
1987.	Habitat selection by breeding Belted Kingfishers (*Ceryle alcyon*). Amer. Midland Natur. 117:63-70.

CARUTHERS, J. H.
1986.	Behavioral and ecological correlates of interference competition among some Hawaiian Drepanidinae. Auk 103:564-574.

Literature Cited

CASWELL, H.
1976. Community structure: A neutral model analysis. Ecol. Monogr. 46:327-354.

CEZILLY, F. and J. WALLACE.
1988. The determination of prey captured by birds through direct field observations: a test of the method. Colonial Waterbirds 11:110-112.

CHAPIN, J. P.
1939. The birds of the Belgian Congo, part II. Bull. Amer. Mus. Nat. Hist. 75:1-632.

CODY, M. L.
1974. Competition and the structure of bird communities. Princeton Univ. Press, Princeton, New Jersey.

COLWELL, R. K., and D. J. FUTUYMA.
1971. On the measurement of niche breadth and overlap. Ecology 52:567-576.

CONNELL, J. H.
1983. On the prevalence and relative importance of interspecific competition: evidence from field experiments. Amer. Natur. 122:661-696.

DAVIS, W. J.
1982. Territory size in *Megaceryle alcyon* along a stream habitat. Auk 99:353-362.

DIAMOND, A. W.
1984. Feeding overlap in some tropical and temperate seabird communities, pp. 24-46 in Schreiber, R. W., ed., Tropical seabird biology. Studies in Avian Biol. No. 8.

DIAMOND, J. M., and J. TERBORGH.
1967. Observations on bird distribution and feeding assemblages along the Río Callaría, Department of Loreto, Peru. Wilson Bull. 79:273-282.

DICKEY, D. and A. J. VAN ROSSEM.
1938. The birds of El Salvador. Field Mus. Nat. Hist., Zool. Ser., No. 23.

DIN, N. A., and S. K. ELTRINGHAM.
1974. Ecological separation between White and Pink-backed pelicans in the Ruwenzori National Park, Uganda. Ibis 116:28-43.

DOORNBOS, G.
1979.   Winter food habits of Smew (*Mergus albellus* L.) on Lake Yssel, the Netherlands: species and size selection in relation to fish stocks. Ardea 67:42-48.

1984.   Piscivorous birds on the saline Lake Grevelingen, The Netherlands: abundance, prey selection, and annual food consumption. Neth. J. Sea Res. 18:457-479.

DORWARD, D. F.
1962.   Comparative biology of the White Booby and the Brown Booby *Sula* spp. at Ascension. Ibis 103:174-220.

DOUTHWAITE, R. J.
1971.   Treatment of fish by the Pied Kingfisher *Ceryle rudis*. Ibis 113:526-529.

1976.   Fishing techniques and food of the Pied Kingfisher on Lake Victoria in Uganda. Ostrich 47:153-160.

DRAULANS, D., J. VAN VESSERN, and E. COENEN.
1981.   Note on piscivorous birds around Lake Kivu, Rwanda. Gerfaut 71:443-455.

EADIE, J. M. and A. KEAST.
1982.   Do goldeneyes and perch compete for food? Oecologia 55:225-230.

ERIKSSON, M. O. G.
1979.   Competition between freshwater fish and goldeneyes *Bucephala clangula* (L.) for common prey. Oecologia 41:98-107.

FEINSINGER, P., E. E. SPEARS, and R. W. POOLE.
1981.   A simple measure of niche breadth. Ecology 62:27-32.

FFRENCH, R.
1973.   A guide to the birds of Trinidad and Tobago. Livingston, Wynnewood, Pennsylvania.

FITZPATRICK, J. W.
1980.   Foraging behavior of Neotropical tyrant flycatchers. Condor 82:43-57.

FORSHAW, J. M.
1983.   Kingfishers and related birds. Alcedinidae. *Ceryle* to *Cittura*. Landsdowne Editions, Melbourne.

Literature Cited

FRY, C. H.
1980a. The evolutionary biology of kingfishers (Alcedinidae). Living Bird 18:113-160.

1980b. Survival and longevity among tropical land birds. Proc. IV Pan-Afr. Ornith. Congress: 333-343.

GALES, R. P.
1988. The use of otoliths as indicators of Little Penguin *Eudyptula minor* diet. Ibis 130:418-426.

GLASSER, J. W., and H. J. PRICE.
1988. Evaluating expectations deduced from explicit hypotheses about mechanisms of competition. Oikos 51:57-70.

GOSS-CUSTARD, J. D., J. T. CRAYFORD, J. T. BOATES, and S. E. A. LEV. DIT DURELL.
1987. Field tests of the accuracy of estimating prey size from bill length in oystercatchers, Haematopus ostralegus, eating mussels, *Mytilus edulis*. Anim. Behav. 35: 1078-1083.

GOULDING, M.
1980. The fishes and the forest. Univ. of Calif. Press, Berkeley.

GRANT, P. R., and I. ABBOTT.
1980. Interspecific competition, island biogeography, and null hypotheses. Evolution 34:332-341.

GREENBERG, R. and J. GRADWOHL.
1986. Constant density and stable territoriality in some tropical insectivorous birds. Oecologia 69:618-625.

GREENE, H. M., G. M. BURGHARDT, B. A. DUNCAN, and A. S. RAND.
1978. Predation and defensive behavior of green iguanas (Reptilia, Lacertilia, Iguanidae). J. Herpetol. 12:169-176.

GRINNELL, J.
1914. An account of the mammals and birds of the lower Colorado Valley. Univ. Calif. Publ. Zool. 12:51-294.

HÄKKINEN, I.
1978. Diet of the Osprey *Pandion haliaetus* in Finland. Ornis Scand. 9:111-116.

HANSKI, I.
1978. Some comments on the measurement of niche metrics. Ecology 59:168-174.

HARDIN, G.
1960. The competitive exclusion principle. Science 131:1292-1297.

HÄRKÖNEN, T. J.
1988. Food-habitat relationship of Harbour Seals and Black Cormorants in Skagerrak and Kattegat. J. Zool. (London) 214:673-681.

HARRIS, M. P.
1970. Differences in the diet of British auks. Ibis 112:540-541.

HAVERSCHMIDT, F.
1968. Birds of Surinam. Oliver and Boyd, Edinburgh.

HESPENHEIDE, H. A.
1973. Ecological inferences from morphological data. Ann. Rev. Ecol. Syst. 4:213-230.

1975. Prey characteristics and predator niche width, pp. 158-180 in M. L. Cody and J. M. Diamond, eds., Ecology and evolution of communities. Harvard Univ. Press, Cambridge, Massachusetts.

HILTY, S. L., and W. L. BROWN
1986. A guide to the birds of Colombia. Princeton Univ. Press, Princeton, New Jersey.

HOLMES, R. T., R. E. BONNEY, JR., and S. W. PACALA.
1979. Guild structure of the Hubbard Brook bird community: a multivariate approach. Ecology 60:512-520.

HOLMES, R. T. and H. F. RECHER.
1986. Determinants of guild structure in forest bird communities: an intercontinental comparison. Condor 88:427-439.

HOLMES, R. T., and S. K. ROBINSON.
1981. Tree species preference of foraging insectivorous birds in a northern hardwoods forest. Oecologia 48:31-35.

HOLT, R. D.
1987. On the relation between niche overlap and competition: the effect of incommensurable niche dimensions. Oikos 48:110-114.

Literature Cited

HOM, C. W.
1983.   Foraging ecology of herons in a southern San Francisco Bay salt marsh. Colonial Waterbirds 6:37-44.

HULSMAN, K.
1981.   Width of gape as a determinant of size of prey eaten by terns. Emu 81:29-32.

1987.   Resource partitioning among sympatric species of tern. Ardea 75:255-262.

HURLBERT, S. H.
1978.   The measurement of niche overlap and some relatives. Ecology 59:67-77.

1982.   Notes on measurement of niche overlap. Ecology 63:252-253.

HUTCHINSON, G. E.
1957.   Concluding remarks. Cold Spring Harbor Symp. Quant. Biol. 22:415-427.

1965.   The ecological theater and the evolutionary play. Yale Univ. Press, New Haven, Connecticut.

JAMES, F. C. and C. E. McCULLOCH.
1985.   Data analysis and the design of experiments in ornithology. Curr. Ornith. 2:1-63.

JANES, S. W.
1984.   Influences of territory composition and interspecific competition on Red-tailed Hawk reproductive success. Ecology 65:862-870.

JOHNSON, N. K.
1975.   Controls of number of bird species on montane islands in the Great Basin. Evolution 29:545-567.

JOHNSTON, D. W.
1989.   Feeding ecology of Pied Kingfishers on Lake Malawi, Africa. Biotropica 21:275-277.

JUNK, W.
1970.   Investigations on the ecology and production-biology of the "floating meadows" (*Paspalo-Echinochloetum*) on the Middle Amazon, part 1: The floating vegetation and its ecology. Amazoniana 2:499-495.

1973. Investigations of the ecology and production-biology of the "floating meadows" (*Paspalo-Echinochloetum*) on the Middle Amazon, part 2: The aquatic fauna in the root zone of floating vegetation. Amazoniana 4:9-102.

JUNOR, F. J. R.
1972. Offshore fishing by the Pied Kingfisher *Ceryle rudis* at Lake Kariba. Ostrich 43:185.

KARR, J. R.
1971. Structure of avian communities in selected Panama and Illinois habitats. Ecol. Monogr. 41:207-223.

1975. Production, energy pathways, and community diversity in forest birds, pp. 161-176 in F. B. Golley and E. Medina, eds.,Tropical ecological systems. Springer-Verlag, New York.

1976. Within- and between-habitat avian diversity in Africa and Neotropical lowland habitats. Ecol. Monogr. 46:457-481.

1980. Geographical variation in the avifaunas of tropical forest undergrowth. Auk 97:283-298.

KARR, J. R. and R. R. ROTH.
1971. Vegetation structure and avian diversity in several New World areas. Amer. Natur. 105:423-436.

KLOPFER, P. H., and R. H. MacARTHUR.
1961. On the causes of tropical species diversity: niche overlap. Amer. Natur. 95:223-226.

KNOPF, F. L. and J. L. KENNEDY.
1981. Differential predation by two species of piscivorous birds. Wilson Bull. 93:554-557.

KUSHLAN, J. A.
1974. Effects of a natural fish kill on the water quality, plankton, and fish population of a pond in the Big Cypress Swamp, Florida. Trans. Amer. Fish. Soc. 103:235-243.

1976a. Environmental stability and fish community diversity. Ecology 57:821-825.

1976b. Feeding behavior of North American herons. Auk 93:86-94.

KUSHLAN, J. A., G. MORALES, and P. C. FROHRING.
1985.   Foraging niche relations of wading birds in tropical wet savannas. Pp. 663-682 in P. A. Buckley et al., eds., Neotropical Ornithology Ornithol. Monogr. No. 36.

LACK, D.
1971.   Ecological isolation in birds. Harvard Univ. Press, Cambridge, Mass.

LEMMETYINEN, R.
1973.   Feeding ecology of *Sterna paradisaea* Pontopp. and *S. hirundo* L. in the archipelago of southwestern Finland. Ann. Zool. Fenn. 10:507-525.

LEVINS, R.
1968.   Evolution in changing environments. Princeton Univ. Press, Princeton, New Jersey.

LINTON, L. R., R. W. DAVIES, and F. J. WRONA.
1981.   Resource utilization indices: an assessment. J. Anim Ecol. 50:283-292.

LOFTIN, H.
1975.   Recaptures and recoveries of banded native Panamanian birds. Bird-Banding 46:19-27.

LOISELLE, B. A. and W. G. HOPPES.
1983.   Nest predation in insular and mainland lowland rainforest in Panama. Condor 85:93-95.

LOWE-McCONNELL, R. H.
1975.   Fish communities in tropical freshwaters. Longman, London.

MacARTHUR, R. H.
1958.   Population ecology of some warblers of northeastern coniferous forests. Ecology 39:599-619.

1964.   Environmental factors affecting bird species diversity. Amer. Natur. 98:387-397.

1965.   Patterns of species diversity. Biol. Rev. 40:510-533.

1972.   Geographical ecology: Pattern in the distribution of species. Harper and Row, New York.

MacARTHUR, R. H. and R. LEVINS.
1967.   The limiting similarity, convergence, and divergence of coexisting species. Amer. Natur. 101:377-385.

MacARTHUR, R. H., J. MacARTHUR, and J. PREER.
1962.   On bird species diversity. II. Prediction of bird census from habitat measurements. Amer. Natur. 96:167-174.

MAURER, B. A.
1982.   Statistical inference for MacArthur-Levins niche overlap. Ecology 63:1712-1729.

MAY, R. M.
1973.   Stability and complexity in model ecosystems. Princeton Univ. Press, Princeton, New Jersey.

1975.   Some notes on estimating the competition matrix. Ecology 56:737-741.

M'CLOSKEY, R. T.
1978.   Niche separation and assembly in four species of Sonoran desert rodents. Amer. Natur. 112:683-694.

MEYER DE SCHAUENSEE, R.
1970.   A guide to the birds of South America. Livingston, Wynnewood, Pennsylvania.

MEYERRIECKS, A. J.
1960.   Comparative breeding behavior of four species of North American herons. Publ. Nuttall Ornithol. Club no. 2.

MONTGOMERIE, R. D., and C. L. GASS.
1981.   Energy limitation of hummingbird populations in tropical and temperate communities. Oecologia 50:162-165.

MORALES, G., J. PINOWSKI, J. PACHECO, M. MADRIZ, and F. GÓMEZ.
1981.   Densidades poblacionales, flujo de energía y hábitos alimentarios de las aves ic tiofagas de los módulos de Apure, Venezuela. Acta Biol. Venez. 11:1-45.

MOURER-CHAUVIRÉ, C.
1982.   Les oiseaux fossiles des Phosphorites du Quercy (Eocene Supérior a Oligocene Supérior): implications paléobiogeographiques. Gèobios Mem. Spec. 6:413-426.

Literature Cited

OBERHOLSER, H. C. and E. B. KINCAID, JR.
1974.   The bird life of Texas. University of Texas Press, Austin.

ONIKI, Y.
1979.   Is nesting success of birds low in the tropics? Biotropica 11:60-69.

ORIANS, G. H.
1969.   The number of bird species in some tropical forests. Ecology 50:783-801.

PEARSON, D. L.
1975.   Un estudio de las aves de Tumi Chucua, departamento del Beni, Bolivia. Pumapunku 8:50-56.

1977.   A pantropical comparison of bird community structure on six lowland forest sites. Condor 79:232-244.

PETRAITIS, P. S.
1979.   Likelihood measures of niche breadth and overlap. Ecology 62:545-548.

PIANKA, E. R.
1966.   Latitudinal gradients in species diversity: a review of concepts. Amer. Natur. 100:33-46.

1972. r and K selection or b and d selection? Amer. Natur. 106:581-588.

1974.   Niche overlap and diffuse competition. Proc. Nat. Acad. Sci. USA 71:2141-2145.

PILON, C., J. BURTON, and R. McNEIL.
1983.   Summer food of the Great and Double-crested cormorants on the Magdalen Islands, Quebec. Can. J. Zool. 61:2733-2739.

REMSEN, J. V., JR.
1978.   Geographical ecology of Neotropical kingfishers. Ph.D. thesis, Univ. of Calif., Berkeley, 207 p.

1985.   Community organization and ecology of birds of high elevation humid forest of the Bolivian Andes, pp. 733-756 in P. A. Buckley et al., eds., Neotropical Ornithology. Ornithol. Monogr. No. 36.

1986.   Aves de una localidad en la sabana húmeda del norte de Bolivia. Ecología en Bolivia 8:21-36.

REMSEN, J. V., JR. and T. A. PARKER III.
1983. Contribution of river-created habitats to bird species richness in Amazonia. Biotropica 15:223-231.

1984. Arboreal dead-leaf-searching birds of the neotropics. Condor 86:36-41.

RICKLEFS, R. E. and G. W. COX.
1977. Morphological similarity and ecological overlap among passerine birds on St. Kitts, British West Indies. Oikos 29:60-66.

RICKLEFS, R. E. and M. LAU.
1980. Bias and dispersion of overlap indices: results of some Monte Carlo simulations. Ecology 61:1019-1024.

RICKLEFS, R. E. and J. TRAVIS.
1980. A morphological approach to the study of avian community organization. Auk 97:321-338.

RIDGELY, R. S.
1976. A guide to the birds of Panama. Princeton Univ. Press, Princeton, New Jersey.

ROBERTS, T. R.
1973. Ecology of fishes in the Amazon and Congo Basins, pp. 239-234 in B. J. Meggers, E. S. Ayensu, and W. D. Duckworth, eds., Tropical ecosystems in Africa and South America: A comparative review. Smithsonian Inst. Press, Washington, D. C.

ROBINSON, S. K.
1981. Ecological relations and social interactions of Philadelphia and Red-eyed vireos. Condor 83:16-26.

ROBINSON, S. K. and R. T. HOLMES.
1984. Effects of plant species and foliage structure on the foraging behavior of forest birds. Auk 101:672-684.

ROOT, R. B.
1967. The niche exploitation pattern of the Blue-gray Gnatcatcher. Ecol. Monogr. 37:317-350.

ROSENBERG, K. V., R. D. OHMART, W. C. HUNTER, and B. W. ANDERSON.
(in press). Birds of the Lower Colorado River Valley. Univ. of Arizona Press, Tucson, Arizona.

RUSTERHOLZ, K. A.
1981a. Niche overlap among foliage-gleaning birds: support for Pianka's niche-overlap hypothesis. Amer. Natur. 117:395-399.

1981b. Competition and the structure of an avian foraging guild. Amer. Natur. 118:173-190.

SCHOENER, T. W.
1970. Nonsynchronous spatial overlap of lizards in patchy habitats. Ecology 51:408-418.

1971. Large-billed insectivorous birds: A precipitous diversity gradient. Condor 73:154-161.

1974. Resource partitioning in ecological communities. Science 185:27-39.

1983. Field experiments on interspecific competition. Amer. Natur. 122:240-285.

SKUTCH, A. F.
1957. Life history of the Amazon Kingfisher. Condor 59:217-229.

1972. Ringed Kingfisher. Pp. 88-101 in Studies of tropical American birds. Publ. Nuttall Ornith. Club no. 10.

1985. Clutch size, nesting success, and predation on nests of neotropical birds, reviewed, pp. 575-594 in Buckley, P. A. et al., eds., Neotropical Ornithology. Ornith. Monogr. no. 36.

SLOBODCHIKOFF, C. N. and W. C. SCHULZ.
1980. Measures of niche overlap. Ecology 61:1051-1055.

SLOBODKIN, L. B.
1962. Growth and regulation of animal populations. Holt, Rinehart, and Winston, New York.

SLUD, P.
1964. The birds of Costa Rica. Bull. Amer. Mus. Nat. Hist. 128:1-430.

SMITH, E. P.
1982. Niche breadth, resource availability, and inference. Ecology 63:1675-1681.

1984. A note on the general likelihood measure of overlap. Ecology 65:323-324.

SMITH, E. P. and T. M. ZARET.
1982.   Bias in estimating niche overlap. Ecology 63:1248-1253.

SNOW, D. W. and A. LILL
1974.   Longevity records for some neotropical land birds. Condor 76:262-267.

STILES, E. W.
1978.   Avian communities in temperate and tropical alder forests. Condor 80:276-284.

STILES, F. G. and A. F. SKUTCH.
1989.   A guide to the birds of Costa Rica. Cornell Univ. Press, Ithaca, New York.

STRONG, D. R., Jr., L. A. SZYSKA, and D. S. SIMBERLOFF.
1979.   Tests of community-wide character displacement against null hypotheses. Evolution 33:897-913.

SWENSON, J. E.
1978.   Prey and foraging behavior of Ospreys on Yellowstone Lake, Wyoming. J. Wildlife Manag. 42:87-90.

1979.   The relationship between prey species ecology and dive success in Ospreys. Auk 96:408-412.

SWING, C. K. and J. S. RAMSEY.
1984.   Lista preliminar de los peces del Lago Tumi Chucua, Provincia Vacadiez, Departamento de Beni. Ecología en Bolivia 5:73-82.

1987.   Preferencias de habitat de los peces del Lago Tumi Chucua, Departamento del Beni, Bolivia. Ecología en Bolivia 10:33-35.

TERBORGH, J.
1980.   Causes of tropical species diversity. Acta XVII Congr. Intern. Ornith.: 955-961.

TERBORGH, J. and J. S. WESKE.
1969.   Colonization of secondary habitats by Peruvian birds. Ecology 50:765-782.

TERBORGH, J. W., J. W. FITZPATRICK, and L. EMMONS.
1984.   Annotated checklist of bird and mammal species of Cocha Cashu Biological Station, Manu National Park, Peru. Fieldiana (Zool.) No. 21.

Literature Cited

THOMAS, B. T.
1979. The birds of a ranch in the Venezuelan llanos, pp. 213-232 in J. F. Eisenberg, ed., Vertebrate ecology in the northern tropics. Smithsonian Inst., Washington, D. C.

THOMSON, J. D. and K. A. RUSTERHOLZ.
1982. Overlap summary indices and the detection of community structure. Ecology 63:274-277.

TJOMLID, S. A.
1973. Food preferences and feeding habits of the Pied Kingfisher *Ceryle rudis*. Ornis Scand. 4:145-151.

WAITE, R. K.
1984. Sympatric corvids. Effects of social behavior, aggression, and avoidance in feeding. Behav. Ecol. Sociobiol. 15:55-60.

WHITFIELD, A. K. and S. J. M. BLABER.
1978. Feeding ecology of piscivorous birds at Lake St. Lucia, part 1: diving birds. Ostrich 49:185-198.

1979. Feeding ecology of piscivorous birds at Lake St. Lucia, part 2: wading birds. Ostrich 50:1-9.

WIENS, J. A.
1977. On competition and variable environments. Amer. Scientist 65:590-597.

1982. On size ratios and sequences in ecological communities: are there no rules? Ann. Zool. Fenn. 19:297-308.

1984. On understanding a non-equilibrium world: myth and reality in community patterns and processes, pp. 439-457 in D. A. Strong et al., eds., Ecological communities. Conceptual issues and the evidence. Princeton Univ. Press, Princeton, New Jersey.

WILLARD, D. E.
1977. The feeding ecology and behavior of five species of herons in southeastern New Jersey. Condor 79:462-470.

1985. Comparative feeding ecology of twenty-two tropical piscivores, pp. 788-797 in P. A. Buckley et al., eds., Neotropical Ornithology, Ornithol. Monogr. No. 36.

WILLIAMS, J. B. and G. O. BATZLI.
1979.   Competition among bark-foraging birds in central Illinois: experimental evidence. Condor 81:122-132.

WILLIS, E. O.
1966.   Interspecific competition and the foraging behavior of Plain-brown Woodcreepers. Ecology 47:667-671.

WILLIS, E. O. and E. EISENMANN.
1979.   A revised list of the birds of Barro Colorado Island, Panama. Smithsonian Contrib. Zool. no. 291.

WILSON, D. S.
1975.   The adequacy of body size as a niche difference. Amer. Natur. 109:769-784.

YOUNG, C. G.
1929.   A contribution to the ornithology of the coastland of British Guiana, part 2. Ibis (ser. 12) 5:1-38.

ZOTTOLI, S. J.
1976.   Feeding segregation in the Arctic and Common terns in southern Finland. Auk 93:636-642.